DuPont: From the Banks of the Brandywine to Miracles of Science

BY ADRIAN KINNANE

Printed in the United States of America
246897531
E. I. du Pont de Nemours and Company
1007 Market Street
Wilmington, Delaware 19898
ISBN 0-8018-7059-3
Library of Congress Control Number 20010988888

TABLE OF CONTENTS

FOREWORD

For a company that has reached its 200th year, transformed itself several times, made inventions that helped shape the modern world and literally become part of the daily lives of millions around the world, DuPont has had relatively few book-length treatments of its history. The most important historical works dealing with the company are scholarly in nature. Although invaluable to the student of DuPont who desires a close reading of particular topics in the company's history, they are less accessible to the general reader. Meanwhile many popular books about the company are either undependable from the historian's perspective, polemical in their aims, or dated.

The present volume was commissioned in an effort to occupy the middle ground: a book researched and written by professional historians, but still eminently readable by employees, retirees and the general public. We believe that author Adrian Kinnane of History Associates Incorporated has succeeded. Moreover, his book traverses familiar terrain in the first several chapters, but in Chapters 7, 8 and 9 offers an overview of the recent history of the company — the past three decades — here presented for the first time in any book. Another objective from the outset was to publish in this volume some of the rich legacy of photographs pertaining to DuPont from the collection of Hagley Museum and Library. Readers familiar with the company will see time-honored images indispensable to telling the company's story, but many other illustrations are less well known.

We are indebted to others at HAI for their contributions: Sarah Leavitt, who assisted in selecting pictures and researching sidebar topics; Ken Durr and Jim Wallace who edited the text; and John Harper who researched photographs; also Gail Mathews, Carol Spielman, and Mary Ann FitzGerald, of HAI staff. We are grateful for the cooperation and guidance of many of the Hagley staff, in particular Marjorie McNinch of the Manuscripts & Archives Department and Jon M. Williams and Barbara Hall of the Pictorial Collections Department. We thank Joel Adler, Carol Adler, Susan Jones and Trish Moore of Adler Design Group, who created the format for the book and who were tireless in their effort and attention to detail. Jeanne Dyson was our very capable production supervisor. Kim Clark coordinated distribution. We also appreciate the copy editing of Janet Macnamara-Barnett and the index by Jan Moore.

The manuscript was reviewed and commented upon by four readers whose judgment, knowledge and insight we regard highly: Geoffrey Gamble of DuPont Legal (and the fifth generation of his family to work for the company); Thomas R. Keane, DuPont Fellow and retired from DuPont Engineering; Daniel T. Muir, Deputy Director for Museum Administration at Hagley; and Richard J. Woodward, retired from DuPont Public Affairs.

To avoid disconcerting readers, we note in advance a few editorial decisions regarding names. While the founder of the company was called Irénée by his family and friends, he more recently has come to be called "E. I." among employees and others close to the company. We have settled on that contemporary convention for this book. The company is properly referred to as DuPont or the DuPont Company, i.e., without a space between the two words in the family name and with an upper case "D" and "P". The family name when used in reference to a family member, is, according to their own practice, presented as "du Pont." The singular exception is the name of Samuel Francis Du Pont who chose a different style; we saw no reason to contradict the admiral.

This book stands as one of the central elements in the commemoration of the 200th Anniversary of DuPont. We hope it informs and delights readers for many years to come.

Justin Carisio, Executive Editor

James Moore, Managing Editor

WILMINGTON, DELAWARE

CHAPTER 1

A VISION AND PRODUCT

TOP LEFT: Family patriarch Pierre Samuel du Pont de Nemours, as painted by Rembrandt Peale in 1810.

TOP RIGHT: Eleuthère Irénée du Pont at age 20. His name means "freedom" and "peace."

RIGHT: More than a dozen members of the du Pont family emigrated to America aboard *American Eagle.* During the grueling voyage, Pierre Samuel kept spirits up — as he had in La Force prison in Paris — by leading others in songs, games and debates.

BELOW: Bois des Fosses, Chevannes. The du Pont family home in France.

RIGHT: The 1801 meeting between (l to r) E. I. du Pont, Paul Revere Franklin, Marquis de Lafayette and Thomas Jefferson, at which Jefferson encouraged E. I. to establish a gunpowder mill.

On Monday, September 17, 1787, a Quaker farmer named Jacob Broom signed the new U.S. Constitution for the state of Delaware. Nearly 4,000 miles away in Paris, Eleuthère Irénée du Pont waited to begin training at his government's gunpowder agency. Whatever the 16-year-old du Pont's youthful dreams of the future may have been, he could hardly have imagined that a place called Delaware would one day be his home, or that Broom would make his settling there so difficult. But in the middle of his training, just as America settled into its new unity, France erupted into violent revolution and du Pont's life was changed forever.

At first the change was small. Eleuthère Irénée (E.I.) completed another year in the national powder works at Essonnes, near Paris, then left to join his father's new printing business in the city. The du Pont family had hoped that the revolution in 1789 would lead to some badly needed political and social reforms. However, the movement soon grew into a harvest of victims for the guillotine. In 1799 Eleuthère's outspoken father, Pierre Samuel du Pont de Nemours, was imprisoned and sentenced to death. But his accuser, Robespierre, met the guillotine first and the elder du Pont was released. Still, France remained a volatile and dangerous place. Pierre Samuel and his son E.I. sold their besieged printing business, gathered their families, and set sail for America on Captain Brooks's *American Eagle*.

Despite its proud name, the vessel was barely seaworthy. The rotting, barnacle-encrusted *American Eagle,* bound first for New York but finally for any port she could reach, made slow headway, weathering 90 days of Atlantic winter storms. Provisions had spoiled in the leaky hold, and the starving passengers and crew had been forced to beg emergency supplies from ships crossing their path. Captain Brooks coaxed his battered ship into harbor at Block Island, off the Rhode Island coast, on New Year's Day 1800. The shivering du Ponts slogged ashore to search for a meal, but there was none to be found. So the family — desperate but still dignified — found food in an empty house and left a gold coin behind. It was the du Ponts' first business transaction on American soil.[1]

Three days later the ship reached the mainland at Newport. Like many Europeans, the elder du Pont believed America to be a land of new hopes and riches as well as a safe haven from the Old World's troubles.[2] Working with some well-placed friends in Paris, he had formed a joint stock company for a land settlement venture in America two years before he and his family left France. Pierre Samuel also had a more personal tie to America. As a highly regarded free-trade advocate, journalist and government official, he had helped Benjamin Franklin negotiate the terms of the Treaty of Paris that ended the American Revolution in 1783. Thomas Jefferson, appointed American envoy to France in 1785, also had befriended du Pont de Nemours and had warned him away from the cutthroat real estate speculation that was spreading across the American West.

The du Ponts found temporary housing in Bergen, New Jersey, and weighed their options. The elder du Pont broached elaborate schemes for international trade, but E.I.'s more practical suggestion was persuasive: The du Ponts should manufacture gunpowder. After all, E.I. had studied at the national Powder and Saltpeter Administration under the direction of his father's good friend, the pioneering chemist Antoine Lavoisier. He had then worked with the latest powder-making technologies during his apprenticeship at Essonnes. Even more important, he had absorbed the spirit of innovation that guided Lavoisier's research. For E. I., black powder was not only a viable product but an evolving technology as well.

In November 1800, E. I. and his friend Major Louis de Tousard visited the Lane-Decatur powder works outside Philadelphia to assess the methods used at one of America's leading powder mills. Tousard was well known there. He had lost an arm fighting for the American cause in the Revolutionary War and now supervised military gunpowder purchases as inspector of artillery for the U.S. Army. The visit confirmed what Tousard probably had told E. I. — that American-made powder could stand improvement.[3] E. I. found fault with nearly every aspect of Lane-Decatur's operations, from the refining of the sulfur and potassium nitrate (saltpeter) to the mixing of those ingredients with charcoal; from the pressing, sifting, drying and glazing of the mixture to the company's management of its workers. "Such competitors should not be formidable," E. I. assured his father.[4] If the competition looked easy, business appeared good. Despite their inefficiencies, William Lane and Stephen Decatur were still turning a sizeable profit. Political conditions in the new land also looked favorable. Alexander Hamilton's 1791 *Report on Manufactures* had specifically urged the domestic production of black powder.

E. I. and his older brother, Victor, headed back to France to raise capital and purchase equipment for the powder mills. They secured from their father's

L3 - 1312

SHIPPED, *in good order and well-conditioned, by* M. Irenee duPont
on board the good Schooner *called the* Betzy of Paterson
whereof is Master for this present voyage, Thos. Van name
now laying in the port of New-York, *and bound for* Wilmington Delaware
To say,

I. D. P. — 1 a 19 & 28 a 31 twenty three Boxes, furniture, Wearing apparel, tools & books
20 a 23 four Boxes
23 a 27 one hogshead & three barrels } tools & implements
32 a 38 Seven trunks, Wearing apparel

C — Eight Barrels Corn & one Barrel Vinegar

no mark — one ten plate Stove — 1 Cart — 1 plough, 1 harrow, two Wheel barrow, 1 hand barrow & 1 chair

Seven Sheeps

being marked and numbered as in the margin, and are to be delivered in the like good order and well-conditioned at the aforesaid port of Wilmington *(the danger of the seas only excepted) unto* M. William Hamon *or to* his *assigns, he or they paying freight for the said* — one hundred Dollars

with ~~primage and average~~ accustomed. In Witness whereof, the Master or Purser of the said Schooner *hath affirmed to* two *Bills of Lading, all of this tenor and date; the one of which* two *Bills being accomplished, the other to stand void. Dated in* New-York, *the* 9th *day of* July 1802.

Thomas Van Name

OPPOSITE TOP: **E. I.'s friend, Louis de Tousard, fought in the American Revolution, then settled in a French community near Wilmington, Delaware. It is believed that while hunting with Tousard, E. I. realized that American powder was inferior to that being made in France, giving birth to the idea of entering the powder business.**

OPPOSITE BOTTOM: **This Bill of Lading accompanied 23 boxes of du Pont family household goods aboard the *Betzy of Paterson* sailing from New York July 9, 1802, for Wilmington.**

LEFT: **E. I. learned the art of manufacturing gunpowder under the tutelage of chemist Antoine Lavoisier, shown here with his wife in a 1788 painting.**

RIGHT: **To form a company, E. I. du Pont had 18 Articles of Association drawn up on April 21, 1801, with his French and American investors.**

OPPOSITE: **E. I. du Pont and his wife, Sophie Madeline Dalmas. Pierre Samuel had strenuously objected when his son married Sophie on November 26, 1791. But she proved a resourceful, loyal and courageous daughter-in-law, handily winning him over. A trademan's daughter, she disguised herself in peasant clothes and brought her father-in-law food and books every day during his 1794 imprisonment.**

Acte d'association pour l'Etablissement d'une Manufacture
de poudre de guerre et de chasse dans les Etats-Unis d'Amérique.

Les Soussignés Du Pont de Nemours, Père et fils et Cie
de New york – Biderman – Catoire Duquesnoy et Cie – et
Eleuthere Irenée Du Pont, voulant établir une manufacture
de Poudre de guerre et de Chasse dans les Etats Unis d'Amérique
forment entre eux une Société pour cet Etablissement et arrêten
en conséquence les articles Suivants:

Art. 1er

Les fonds de l'entreprise seront de trente six mille
Dollars, formant dixhuit actions de Deux mille Dollars chacun

Art 2.

Les Actions seront fournies par
Biderman pour une action .... une action
Catoire Duquesnoy & Comp. .... une action
Necker Germany .... une actio
Archd. McCall .... { une acti / une acti }
Peter Bauduy .... { une acti / une acti }
Du Pont de Nemours Père et fils et Cie de New york · Douze a

Art. 3.

Les Actions porteront un interet de Six pour cent

Art. 4.

Le Citoyen E. I. Du Pont est chargé d'établir la ma
et d'en diriger les travaux, il y donnera tous ses Soins, e
il lui sera alloué une indemnité annuelle de dixhuit Cen

Art 5.

Les Constructions nécessaires pour l'Etab
de la fabrique seront exécutées pendant le Cours d
1801 et les premiers mois de la Suivante, en s
... 1802.

associates a transfer of $24,000 of the old land settlement company's capital into the new E. I. du Pont de Nemours & Co., which was organized as a *societé en commandité*, or a limited liability, joint stock company in France on April 21, 1801. The brothers planned to raise another $12,000 in America on their return. Equipment, E. I. learned, was easy enough to obtain. Napoleon was eager to undermine Britain's powder export business and more than happy to sell the du Ponts powder-making machinery at cost. Wealthier and wiser, the brothers set sail for America once again, this time on the *Benjamin Franklin*.

E. I., who assumed the title "Director of Manufacturing," faced new problems as soon as he set foot in Philadelphia. First, the joint stock company had to be reorganized as a partnership because American law did not yet recognize the limited liability, private corporation represented by the French *commandité*. But a partnership did not limit shareholders' liability solely to the amount of their investment, so the formal partnership agreement included conditions — then unenforceable under American law — that protected shareholders by keeping them officially anonymous. To ensure that the handful of original shareholders kept control of the company, they were given first option on any stock sales.[5] E. I.'s second problem was deciding where to locate the company's powder mills, and this search led to his first encounter with legendary Yankee shrewdness in the person of Jacob Broom.

At 30 years of age, E. I. was tall and sturdy with a confident disposition. His expression, as captured in midlife by the renowned American portraitist Rembrandt Peale, combined quiet resolve and determination with a sense of introspection, even melancholy. Not as outgoing as either his father or brother Victor, E. I. handled his many responsibilities with an abundance of common sense. His wife, Sophie, shared both his pride and his reserve. Neither of them sought out the hazards of new adventures but both met them with an uncomplaining competence. It had been that way in France during

NOTABLE

> This 1806 drawing of the powder yards by Charles Dalmas — E. I.'s brother-in-law — is the first known visual representation of the DuPont mills. The drawing was sent back to France to reassure stockholders of the company's progress.

> The company's first investors were three Europeans and two Americans, all friends or acquaintances of the du Ponts. Jacques Bidermann was a Swiss banker living in Paris. Louis Necker was the brother of Jacques Necker, ex-finance minister to Louis XVI. Catoire, Duquesnoy & Cie was a Paris bank. Adrien Duquesnoy had been imprisoned with Pierre Samuel du Pont during the French Revolution. Archibald McCall was a prominent Philadelphia merchant and a friend of Peter Bauduy, the other American investor.

> The du Ponts took steps to reassure Americans that they were not violent French revolutionaries. Their choice of Alexander Hamilton, George Washington's Secretary of the Treasury and a leading Federalist, to be their attorney was a shrewd one, though Hamilton did not serve them for very long. He died in 1804 from wounds received in a duel with Vice President Aaron Burr.

> The *American Eagle* carried the du Ponts into a heated political atmosphere in 1800.

The American Revolution ended with peaceful self-government, but the new nation still argued over how power should be balanced between the people and a central government. Federalists like John Adams and anti-Federalists like Thomas Jefferson used both the ideals and the violence of the French Revolution to support their respective views.

> DuPont made its first overseas export of black powder in 1805 to the Spanish government. By 1834, exports accounted for 20 percent of the company's production, with powder shipments being made to Europe, the West Indies and South America.

MEMBERS OF THE DU PONT FAMILY MENTIONED IN THIS BOOK OR WHO ARE KEY TO UNDERSTANDING THE COMPANY'S MANAGEMENT AND SUCCESSION

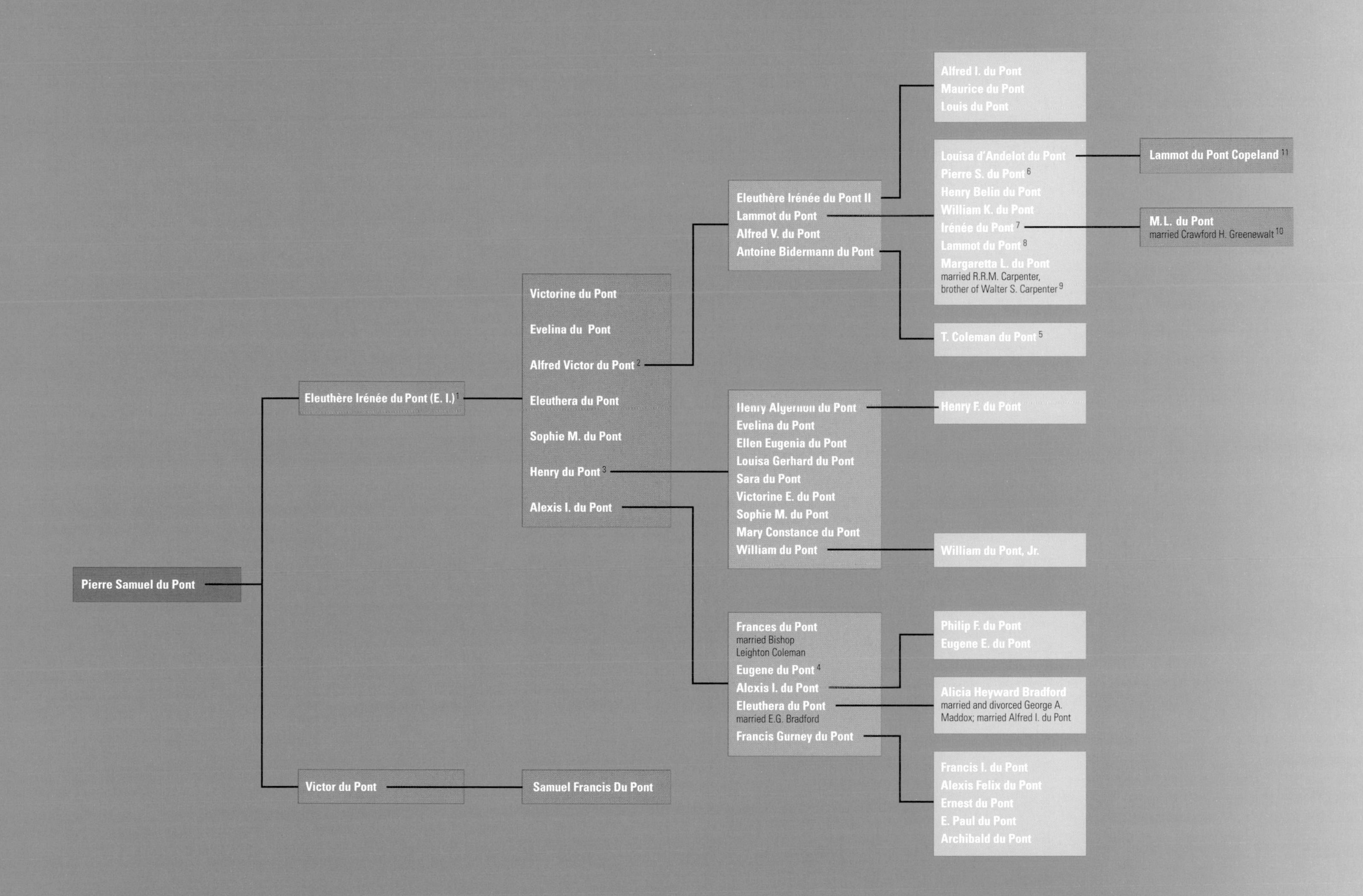

NUMBERS IDENTIFY SUCCESSIVE HEADS OF THE COMPANY.

Draught of a Tract of land of Mr. Dupongt, lately purchased of Jacob Broom Esqr. situated in Christiana Hundred and County of Newcastle in the State of Delaware, bounded and described as on the above draught is set forth and containing Sixty five acres of land,
Surveyed August 13th. Annoque Domini 1802

Pr. Isaac Stevenson

the revolution, and it would be the same with the powder company in America.

In the autumn of 1801 Jacob Broom put all this to the test. E. I. spoke no English and Broom spoke no French, but it was money rather than language that divided them. After signing the Constitution, Broom had returned to Wilmington, where he was a civic and business leader. Broom's cotton mill on the Brandywine River had burned down in 1797, and the 65 acres four miles upstream from Wilmington looked like an ideal place for the du Pont powder mill.

But was it worth Broom's asking price of $7,000? The outlay would consume more than a fifth of du Pont's available capital before construction had even begun. And what was E. I. to make of the rumors that Broom was quietly harvesting some of the property's valuable hardwood trees whose lumber E. I. needed for buildings? The younger du Pont had shouldered a tremendous responsibility in the form of $24,000 — about $240,000 today — invested by his father and his father's friends in France and another $8,000 contributed by two American businessmen. He countered Broom's price with an offer of his own. The property, he declared firmly, was worth $6,000 and no more. The negotiations with Broom quickly stalemated, and E. I.'s anxiety mounted. Where would he go now?

Tousard asked Lane and Decatur on E. I.'s behalf whether they would consider selling their works, but neither man was interested.[6] E. I. had also scouted locations from New York's Catskill Mountains to the new Federal City on the Potomac, but in the end none looked as promising as the lower Brandywine. The river's east and west branches gathered in the streams tumbling down Pennsylvania's Welsh Mountains and herded them into a single course about four miles north of Chadd's Ford, Pennsylvania. The combined waters then raced downhill across the Delaware state line before coasting past Wilmington into the Atlantic-bound shipping lanes of the Delaware River.

Falling water, not steam or electricity, drove the gears of industry in 1800, and the lower five miles of the Brandywine already powered several bustling operations. E. I. surveyed them all — the pounding hammers of the leather, textile and paper factories; the millstones grinding wheat, corn and linseed; even the mill pulverizing tobacco into snuff. The Brandywine's average flow of 4,500 gallons, or 19 tons, per second was powerful enough to drive the machinery of a large mill, survive summer drought and winter freeze, and ensure nearly year-round production. The willow trees abundant on the riverbanks would make excellent charcoal, a key ingredient in black powder. The Broom property also was close to the wharves, yet still far enough from the city to keep workers from taking too easily to other employment or, worse, straying into taverns and drunkenness. So with an émigré friend acting as translator and intermediary, E. I. finally found a crack in his adversary's granite sense of value and pride. On April 27, 1802, he paid the hard-bargaining Broom $6,740 of his precious capital, clearing the way for construction to begin.

Broom must have decided he had sold too cheaply, because he soon tried to force du Pont into another transaction. Broom still owned another tract of land upstream, so he dammed up the river there and reduced du Pont's water supply. His next offer was exasperatingly simple: If E. I. would buy the additional land upstream, he was welcome to tear down the dam. Furious at Broom's arm-twisting, E. I. discovered another parcel of land directly across the river from Broom's property, bought it and, exercising his riparian rights, tore down exactly half of the dam. That move effectively ended any further incursions by Broom on du Pont's budding prosperity.

But it did not end E. I.'s worries. For the remaining 32 years of his life, E. I. brooded endlessly over finances. He was always aware of the many mishaps — and occasional disasters — that could finish a business enterprise in the early 19th century. In an age before insurance, fire or flood could ruin a life's work in a matter of hours, and the funds of family and

Once Isaac Stevenson completed a survey of the du Pont property *(TOP)*, E. I. du Pont drew his plan for the site. The du Ponts chose the site in the Brandywine Valley in Delaware for their powder mills because it was close to wharves for trading yet far enough away from a city to discourage distraction, and the Brandywine River provided power for the mills.

INSET: The first DuPont mills hugged the banks of the Brandywine River about four miles northwest of Wilmington in northern Delaware.

friends were often all that stood between success and failure. The constant danger of explosion made powder making an especially risky business, and E. I. was rarely free of a gnawing sense of imminent catastrophe. But he also knew that a growing America would always need high-quality gunpowder and explosives. So on July 19, 1802, 31-year-old E. I. du Pont set to work on a curving stretch of the Brandywine to make a new start in a country younger than himself.

Two years of toil produced a powder works the du Ponts named Eleutherian Mills. Its dozen buildings with five water wheels included a sawmill for cutting lumber, a storage magazine for finished powder awaiting shipment and housing for the mills' 30 workers and their families. A cooperage on the premises supplied the barrels required for transporting the powder in mule-drawn Conestoga wagons, or "powder schooners," to the Wilmington wharves for shipping.

The design of the du Pont mills perplexed the masons and carpenters E. I. hired to build them. Instead of the single, large building more commonly used in industrial operations, du Pont insisted upon several small ones spaced widely apart. And instead of the solid structures with four thick walls and sturdy roofs that both common sense and tradition demanded for dealing with dangerous materials, E. I. insisted on buildings with only three very thick stone walls and a thin, wooden one facing the river, all covered by a light roof slanting down toward the water. His reasons were grimly sensible. In the event of an accidental explosion, this design would limit the damage, channeling the blast force up and out over the river, away from other buildings and workers. The few men unlucky enough to be caught inside, however, would become human projectiles in the architectural equivalent of a cannon barrel. No wonder that the dour euphemism later used by the workers for dying in an explosion was "going across the creek." Occasionally, that is just what happened.

Since refining of the raw materials came at an early stage in the powder-production process, the potassium nitrate, or saltpeter, refinery at Eleutherian Mills was built first, being completed in the summer of 1803. Saltpeter was until that time obtainable only from British-controlled India, so it alone was salable to governments as a product to be stored for future military needs. With the saltpeter refinery complete, Pierre Samuel and E. I. informed President Thomas Jefferson that they were ready to fill any government contracts for this valuable substance. Earlier that year, the president had asked Pierre Samuel's help in negotiating the Louisiana purchase with Napoleon. Pierre's advice helped the American envoys close the deal. Jefferson now quickly arranged a purchase of saltpeter through the War Department. The following year, when the entire powder plant was operational, Jefferson assured the du Ponts that there would be more business. It came in 1805, when the U.S. Marines ended four years of hostilities with Barbary pirates by storming Tripoli in North Africa to rescue American hostages.

By that time the du Ponts had been making black powder for a year, shipping much of it to retail merchants in 25-pound or 100-pound barrels under the name "Brandywine Powder." The first barrel went to New York in May 1804, consigned to E. I.'s older brother Victor, who promised to do his best to sell it. He didn't have to try very hard. The powder's quality was such that it quickly acquired a top-of-the-line reputation, like Kentucky whiskey or Virginia tobacco. In 1808 a competing Connecticut mill began producing what it called "Brandywine" powder, so the du Ponts capitalized the "d" in their company title and renamed their product "Du Pont Powder."[7]

The quality of black powder was measured simply but effectively: fire two rounds or shots with equal measures of two different powders and see which shot went farther. An instrument called an eprouvette served the same purpose by directing the explosion of a small charge of powder against a spring-loaded gauge. Secretary of War Henry Dearborn ordered that several domestic powders be tested and

TOP: Two drawings of Eleutherian Mills. On the left is Eleuthera du Pont's 1823 sketch of her family's residence. The 1810 brown wash on the right is by the French Baroness Hyde de Neuville, one of the earliest visitors to the mills to leave behind a pictorial record.

The sturdy and versatile Conestoga wagon was the 19th century's truck. Depicted at left in Howard Pyle's 1911 illustration and below, it was invented in the mid-18th century in Pennsylvania. The Conestoga could haul as much as five tons and helped open up new markets for American manufacturers. During the War of 1812, DuPont wagons carried powder over extended distances and difficult terrain to supply Commodore O. H. Perry on Lake Erie. After defeating the British squadron, he reported, "We have met the enemy and they are ours."

DUPONT GUN POWDER
MANUFACTURED BY
DU PONT DE NEMOURS & Co. WILMINGTON, DEL.
BUILT 1802. UPPER HAGLEY MILLS BUILT 1812. LOWER HAGLEY MILLS BUILT 1828. BRANDY WINE MILLS BUILT 1836.
ALSO LUZERNE COUNTY, PENNA.
WAPWALLOPEN MILLS BUILT 1859. GREAT FALLS MILLS BUILT 1869.
Washington August 4th 1803
Sir
From information received from Mr. E. Livingston a few days since, relative to your powder manufactory &c. I wrote to Genl. Irvine, Superintendent of Military Stores at Philadelphia, and requested him (in conformity to what I desired Mr. Livingston to communicate to you) to furnish on your application, such samples of powder and salt petre as you might desire, and to obtain from you the terms on which you would purify
required purific
terms for fi
1826 Aug.
EAGLE GUNPOWDER
RIFLE SHOOTING
DUPONT
WILMINGTON DELAWARE
ESTANCO DE POLVORA
CON
PERMISO DEL GOBIERNO
Polvora Fabricada por
A. F. & Co.
SANTIAGO DE CUBA
DU PONT'S
GUNPOWDER
AND ALL OTHER KINDS.
EAGLE GUNPOWDER
GUNPOWDER FOR ORDNANCE
E. I. DU PONT DE NEMOURS & Co.
Wilmington, Delaware
GRANDE
POLVORA SUPERIOR
PARA MINAS.
Fabricada expresamente para el uso de los Mineros de California y Mexico, por
E. I. Du Pont de Nemours & Co.
R. GIBBONS, San Francisco,
UNICO AGENTE PARA LA COSTA PACIFICO.
POLVORA SUPERIOR

Work done at DuPont's Powder Mills since the beginning of their establishment.

	New Powder made — Eagle Powder	New Powder made — Common Powder	Old Powder remanufactured	Total	amount pr. Sales & Invo Books
1804	"	38525	"	38525	$ 15116
1805	"	77210	75000	152210	46857
1806	"	107219	67200	174419	45400
1807	"	129076	32950	162026	47694
1808	"	144400	93900	198300	53863
1809	"	163006	40300	203306	71183
1810	"	184975	400	185375	866
1811	"	204056	"	204056	122
1812	"	299788	"	299788	14
1813	"	335677	"	335677	21
1814	"	519551	825	520376	29
1815	"	461700	"	461700	2
1816	"	551250	"	551250	1
1817	"	703831	33600	737431	18
1818	"	350896	"	350896	8
1819	"	491964	"	491964	110
1820	"	477179	"	477179	979
1821	"	614086	38625	652711	135836
1822	"	514540	15187	529727	114622
1823	5037	628029	23048	656114	138627
1824	10521	668128	25435	704084	169425
1825	10584	650611	58741	719936	150452
	26,142	8,275,697	505,211	8,807,050	$ 2705629
1826	9,391	569,051	28,072	606,514	127403
	35,533	8,844,748	533,283	9,413,564	$ 2,833,032
1827	11,742	676,620	20,561	708	
	47,275	9,521,368	553,844	10,122	

SUPERIOR SHIPPING GUNPOWDER.

WILMINGTON. DEL.

Caution!

To Dealers in Gun-Powder.

AS there are now different Powder Mills established on the Brandywine creek, the subscribers find it necessary to inform the public and their customers, that to prevent mistakes, they have declined using the name of Brandywine, by which their powder has been heretofore known. In future it will be designated only as Du Pont & Co's. Powder. The kegs and barrels will be marked D. P. & Co.

This powder is easily known from any other by the shape and hardness of its grain. It is warranted equal to any imported, and is far superior to the highest Pennsylvania proof.

Orders for powder of any description, as cannon, musket, F. & FF. glazed or rough, riffle or eagle powder, sent either to Mr. Archibald M'Call, Merchant in Philadelphia, or to the subscribers in Wilmington, shall be duly attended to.

The late extention given to their manufactory will enable the subscribers to execute all orders at the shortest notice, and at reduced prices.

E. J. Du Pont de Nemours & Co

Wilmington, June 11—th1t mwf3mo

A Revolution in the U. States,

" Worth makes the man, the want of it the

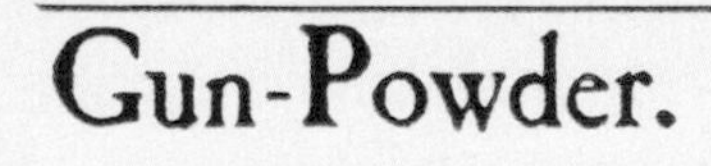

Gun-Powder.

The subscriber offers for sale,

AMERICAN manufactured Gun Powder, from the Brandywine Mills, of a quality which is warranted equal and believed to be superior, to any imported from Europe, and at prices much under those of the imported Powder

The Brandywine Manufactory, which is lately established is upon the most modern and approved plan, being the same that has been adopted by the administration of Gun-Powder in France

From actual experiments, the Powder now offered for sale has been found to be much stronger and quicker than the generality of that which is imported from Europe, and it will be found to resist damp much longer; but as experience is the best guide, persons desirous of purchasing are invited to make a trial, and satisfy themselves.

Orders for Powder of any descriptions, from Cannon to that of the finest grain will be received and duly attended to, by the subscriber who has now on hand, a quantity of Musquet and Rifle Powder, which will be disposed of on moderate term.

Archibald M'Call,

Oct. 19—mwftf No. 187, s. 2d street

LEFT: "Gunpowder. The subscriber offers for sale." This 1804 newspaper clipping is the first ad for DuPont powder, manufactured at "the Brandywine Mills."

Other powder makers followed the du Ponts to the Brandywine and used the same product name, prompting the company to publish a notice (far left) June 11, 1807 — "Caution!" — to powder buyers announcing adoption of the name "Du Pont & Co.'s Powder."

Also shown are early DuPont product labels, ads, sales records, timesheets and company correspondence.

pronounced DuPont's the best. But he was obliged for political reasons to parcel out contracts among various producers, including Stephen Decatur, co-owner of the Lane-Decatur mills and father of Lieutenant Stephen Decatur, hero of Tripoli. DuPont powder surpassed the products of all its American competitors because the company observed strict standards, used the latest technology and production methods and practiced sound management. All these advantages depended on a well-trained and stable work force. Like the physical plant itself, the system of training and retaining employees at the company resulted from E. I.'s practical vision.

The demands of powder making were sharply at odds with what historians have called "pre-industrial" work habits. 18th- and early 19th-century laborers had not yet made the transition from the extended, often slow-paced, seasonal patterns of agriculture to the much faster, machine-like rhythms and clock-induced discipline of factory work. Early industrial workers would often take unannounced leaves of indefinite duration to help with family farming, to celebrate religious holidays or other festivals or simply to enjoy drinking with their friends. They were notoriously, even dangerously, lax about observing carefully developed rules and safety procedures such as those governing powder making. One of the rules E. I. posted on New Year's Day 1811 may have been especially irksome: "All kind of play or disorderly fun is prohibited."[8] Other safety rules were more straightforward, including the requirement that workers wear shoes made with wooden pegs instead of nails and consent to searches of their pockets at the plant gates to keep matches out of the work area.

E. I. du Pont tried to recruit French workers at first. Although it was not difficult to find laborers in America, he confided to a friend that it was hard "to keep the same ones for any length of time."[9] Few French workers were willing to emigrate, however, so E. I. did what he could with the Scots-Irish, Welsh, German, English and Dutch farmers and jack-of-all-trades laborers who gravitated to the Delaware Valley. At first, building a work force entailed trade-offs. Drinking was a case in point. E. I. allowed workers the daily ration of rum that at the time was thought to be nourishing and strength-giving but insisted that the men have it at the end of the day instead of in the middle, when the slightest misstep or careless move could trigger a disaster. But on March 19, 1818, a terrible explosion attributed to a foreman's drinking killed 40 people and injured E. I.'s wife, Sophie, as she played with her toddler in the family home built nearby. That put an end to any compromise on alcohol.

E. I. preferred to hire inexperienced workers and train them in the rules and procedures that he knew would produce superior powder and minimize the risks of injury and death. Unless they had been trained in France, E. I. reasoned, experienced laborers were likely to bring ingrained bad habits to their jobs. Training alone, however, did not solve the problem of retaining skilled and semi-skilled employees. It took several years for E. I. to hammer out the system of reciprocal obligations and shared interests that kept workers in the mills and that came to characterize the DuPont Company's relationship with its employees.[10] By 1811, E. I. had introduced overtime and night pay and offered additional employment to workers' family members. In 1813 he implemented a savings plan offering 6 percent interest on accounts of $100 or more.

An explosion in 1815 had killed nine men — the first casualties at DuPont. But it had also compelled E. I. to establish a pension plan for widows and orphans that assured workers that their families were secure and encouraged them to stay on the job. The du Ponts also shared the plant's dangers, working alongside their employees and showing by example their commitment to the business at every level. If workers' families lived at the mills, so too would the du Ponts. E. I. and Sophie raised seven children in their three-story, stuccoed, stone house overlooking the powder mills. Just across the river, Victor du Pont and his wife, Ann, raised a family of four. When an occasional explosion hit, the du Pont homes suffered the same

Dear Sir

Monticello Apr. 24. 11.

We are, four of us, sportsmen, in my family, amusing ourselves much with our guns. but the powder sold here is wretched, carrying the index of the French eprouvette (such as you furnished Genl. Dearborne) to 9. 10. or 11. only, while the cannister of your powder, recieved from you 2. or 3. years ago, carried it to considerably upwards of 20. I have persuaded a merchant in this neighborhood to get his supply from you which he has promised to do, and I am in hopes the difference which will be found between that & what has been usually bought will induce our other merchants to do the same. I promised mr Lietch, the merchant alluded to, a letter to you when he should go on. this will serve instead of it. but he does not go on till autumn. in the mean time I am engaged in works which require a good deal of rock to be removed with gunpowder, in doing which with the miserable stuff we have here, we make little way. will you be so good as to send me a quarter of a hundred of yours, addressed to Messrs. Gibson & Jefferson of Richmond, who will forward it to me. the cost shall be remitted you as soon as made known. vessels pass from Philadelphia to Richmond almost daily, & the sooner I recieve it, the sooner I shall make effectual progress in my works. Accept the assurances of my great esteem & respect.

Mr. E. Dupont de Nemours

Th: Jefferson

Washington Nov. 23. 04.

Dear Sir

It is with real pleasure I inform you that it is concluded to be for the public interest to apply to your establishment for whatever can be had from that for the use either of the naval or military department. the present is for your private information; you will know it officially by applications from those departments whenever their wants may call for them. Accept my friendly salutations & assurances of esteem & respect.

Th: Jefferson

The relationship between the du Pont family and Thomas Jefferson began when Jefferson served as an American minister to Paris. His business brought him into contact with Pierre Samuel du Pont, a prominent economist with whom he shared an interest in agriculture. E. I. met the minister as a young teenager. In 1800 Jefferson welcomed the arrival of E. I. in America, both for his powder-making skills and out of fondness for his father, Pierre Samuel, who had proved a "faithful and useful friend to this country during my ministry in Paris." Included in the correspondence between E. I. and Jefferson is a letter of November 23, 1804 (*below*), in which Jefferson requested gunpowder for the War Department. In November 1811, Jefferson wrote (*left*) to thank E. I. for the keg of powder he had received, which he "found of very superior kind." The Founding Father continued to request powder both for hunting and for blasting rock at Monticello, his estate in Virginia. Jefferson even acted as a salesman for DuPont, telling E. I., "Having distributed the canisters among the merchants and gentlemen of this quarter, I presume it will occasion calls on you from them."

TOP: The Hagley Yards Sunday School was run by du Pont family members and attended by children of the powder workers.

MIDDLE: Pierre "Peter" Bauduy, part of the French community in Wilmington when E. I. arrived, helped translate for the French-speaking E. I. until he could speak English.

BACKGROUND: The Hagley insurance survey map, done in 1797, suggests the property layout prior to du Pont's purchase in 1812.

BOTTOM: During the War of 1812, Camp DuPont was set up in Montchanin near the DuPont powder works. It served as home for an advance light brigade under Brigadier General Cadwaladar.

OPPOSITE TOP: DuPont, Bauduy, and Company letterhead. E. I. resented Peter Bauduy's insertion of his name on company stationary.

BOTTOM: Alexis Irénée du Pont, youngest son of E. I. and Sophie, was killed in a mill explosion in 1857.

broken windows and cracked walls as the workers' homes.

After the 1818 catastrophe, nearly all the workers fled the mills and refused to come back. The blast had been a powder worker's nightmare, for it had spread from the glazing mill and reached the 30 tons of powder stored in the magazine. Wilmington residents four miles away heard the du Pont magazine thunder into the hillside above the riverbank, then back again to pummel the mills with a hailstorm of stone and dirt. Repairs were made within a month, however, and when E. I. and Victor announced that they would operate the works by themselves if need be, most of the men returned to their jobs. Over the years, as workers handed their steady, well-paying jobs and their DuPont homes down to their sons, employees developed an exceptional loyalty to the firm.

E. I.'s daughters also helped to forge a community at the mills. The eldest, Victorine, taught for most of her life in the Brandywine Manufacturers' Sunday School, incorporated by E. I. in 1817 to teach the three R's, not religion, on the workers' day off as an incentive for the laborers in the days before free public schooling. Her sisters Sophie and Evelina helped at the school over the years and established a temperance society during the 1830s, when religious revivals and reform movements, in which women played a large role, swept the country.[11] Late in that decade, American-born workers began to give way to Irish Catholic immigrants, and the Sunday School's broad-minded ecumenism accommodated the needs of these new mill employees.

Despite the setbacks of explosions and the challenge of maintaining a satisfactory labor force, the young DuPont firm thrived. But this success brought its own set of problems. During the War of 1812, when the company sold half a million pounds of powder to the U.S. government, the French stockholders soon demanded dividend payments from all the profits they imagined to be flowing into company coffers. Yes, there had been profits, E. I. explained, but it had been necessary to use that money to expand the plant to meet the sudden increase in orders. Besides, how would the company fare when the war ended and government sales dropped?

In 1812, E. I. had acquired "Hagley," the Thomas Lea farm just downstream, as a site for additional powder works to meet the wartime demand. E. I.'s $47,000 purchase nearly doubled the size of the DuPont mills, but it had been made despite the protests of his friend Peter Bauduy, who had helped negotiate the deal with Broom and held $8,000 in stock. The company's early years had been a time of almost continuous financial strain, and Bauduy had helped E. I. through it by posting his own property as collateral against DuPont bank loans. He was convinced that his personal investment had earned him some prerogatives. Bauduy's practice of signing business letters "Du Pont, Bauduy & Company" had already strained their relationship, and now Bauduy charged E. I. with concealing profits and diverting shareholder dividends into unnecessary expansions.

In 1813, as the marriage of E. I.'s daughter, Victorine, and Bauduy's son, Ferdinand, approached, the two friends papered over their problems, and when Ferdinand died of pneumonia just a few weeks later, the shared loss momentarily overcame business differences. The conflict soon resurfaced, however. Bauduy's charges that E. I. was squandering profits prompted one of the stockholders, a Swiss banker named Jacques Bidermann, to dispatch his son James Antoine to review the company's books. Young Bidermann found the books in order and the company well-managed, and since E. I.'s sons and likely successors, Alfred, Henry, and Alexis, were still in school, he remained to become E. I.'s assistant. Bidermann joined the family as well, marrying E. I.'s second daughter, Evelina, in 1816. By then DuPont was the largest powder manufacturer in America, and E. I. and Bidermann were able to buy back Bauduy's shares. Bauduy used the money to set up his own "Brandywine Powder" plant on the nearby Christina River and later filed suit over the value of his shares. The matter was finally resolved in 1824 in the DuPont Company's favor.[12]

NOTABLE

> E. I.'s daughters were well qualified to teach school. They had studied art, music, natural science and mathematics at Mademoiselle Rivardi's school in Philadelphia. Like most female students of their social class, they had also practiced useful decorative arts such as needlework. Many graduates went on to establish their own schools. Between 1817 and her death in 1861, Victorine helped to educate nearly 2,000 children and teenagers at the Brandywine Manufacturers' Sunday School.

> Eleuthère Irénée's business interests were not limited to black powder. In 1805 he purchased Don Pedro, a prize ram, for sixty dollars (nearly $700 in 2002 dollars). Don Pedro returned the investment many times over, and E. I. soon had enough livestock to support a woolen mill on the Brandywine. Don Pedro achieved national notoriety among sheep breeders. When he died in 1811, the du Ponts received condolence letters from across the country, most notably from Thomas Jefferson.

> The survivors of the dreadful explosion at the Brandywine Mills on March 12, 1818, were a curious lot. Two workmen, blown some 200 yards, walked away with only bruises. One of them, according to a newspaper account, "was literally blown out of his slippers." Those were later found stuck to the ground where he had been standing. Only one article in E. I. du Pont's house survived intact: a portrait of Napoleon Bonaparte successfully held its ground.

> DuPont implemented its first employee savings plan in 1813 to encourage financial independence among its employees. Under the plan, employees earned 6 percent interest on balances of $100 or more deposited with the company at the end of each year.

In 1818, E. I. du Pont purchased and subdivided a large tract of land in western Pennsylvania to help deserving employees establish their own homesteads. He also arranged for the disbursement of ample credit to his former workers to assist their transition from powder making to farming.

> In 1850, *Scientific American* described DuPont's facilities as "the most extensive powder mills in the world."

> A flood on the Brandywine in 1822 knocked over the walls of the new rolling mill "which were two scaffolds high," according to company records. The DuPont rolling mills — in which large, four-ton, iron wheels ground and mixed the powder, replacing the old up-and-down stamping method — were the first in America. They were phased in over five years. The planned installation in 1822 was delayed until 1824 because the heavy rollers, cast in a foundry in West Point, New York, slid off the ship's deck in a storm while being shipped to Wilmington. Antoine Bidermann selected Michael Callighan, the company's youngest worker, to be DuPont's first "rolling mill man." The new mill blew up just a month later and young Callighan was burned but recovered.

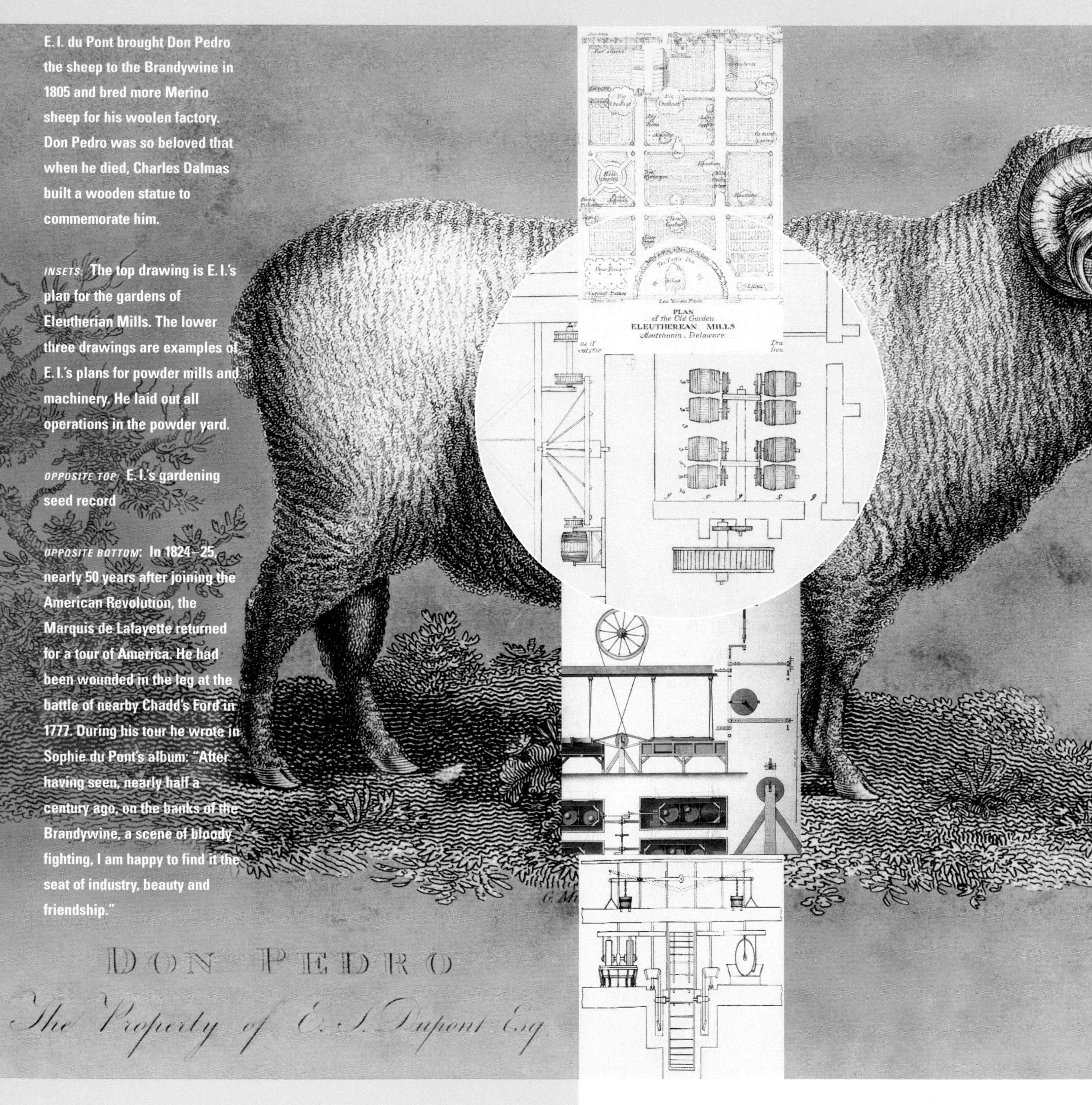

E. I. du Pont brought Don Pedro the sheep to the Brandywine in 1805 and bred more Merino sheep for his woolen factory. Don Pedro was so beloved that when he died, Charles Dalmas built a wooden statue to commemorate him.

INSETS: The top drawing is E. I.'s plan for the gardens of Eleutherian Mills. The lower three drawings are examples of E. I.'s plans for powder mills and machinery. He laid out all operations in the powder yard.

OPPOSITE TOP: E. I.'s gardening seed record

OPPOSITE BOTTOM: In 1824–25, nearly 50 years after joining the American Revolution, the Marquis de Lafayette returned for a tour of America. He had been wounded in the leg at the battle of nearby Chadd's Ford in 1777. During his tour he wrote in Sophie du Pont's album: "After having seen, nearly half a century ago, on the banks of the Brandywine, a scene of bloody fighting, I am happy to find it the seat of industry, beauty and friendship."

The new Hagley Yard allowed the firm to maintain production while damage from the 1818 explosion was being repaired, but the cost of the disaster accounted for much of the company's nearly $200,000 in nonoperating losses incurred between 1817 and 1819. Only E. I.'s reputation for integrity and the demand for his product persuaded the Philadelphia banks to provide the loans needed to keep going. Despite these setbacks, by the 1820s the success of the company seemed assured.

The du Ponts' social success also seemed certain. Victor, who left the running of his woolen mill to son Charles, became a successful Delaware politician. In 1824 Charles married Dorcas Van Dyke, daughter of a U.S. senator. Their wedding in New Castle, Delaware, was the social event of the season, with the Marquis de Lafayette giving the bride away during a visit to America. E. I.'s accomplishments spread beyond the family business as well. In 1814 he became director of the Farmers Bank of the State of Delaware, and in 1822, he was named a director of the Second Bank of the United States.

But even as the du Pont family began to achieve success and recognition in their adopted land, death began to winnow its ranks. Pierre Samuel died in August 1817 at the age of 77, after having exhausted himself participating in a fire brigade that narrowly prevented an explosion in July. Victor died in 1827, and E. I.'s wife passed away the next year. The loss staggered E. I. — Sophie had been a true partner through trying times.

On October 31, 1834, at age 63, E. I. himself collapsed, probably from heart trouble, on the street near his hotel while on business in Philadelphia. His body was brought back to Wilmington by steamboat the next day and laid to rest near his house in the family cemetery at Sand Hole Woods, at the foot of what is now Buck Road. Although he had dedicated his life to his business, E. I. had still found time for civic and charitable causes, including free public education, care of the blind, and provision for the poor citizens of New Castle County. He was widely eulogized as a useful and valuable citizen and as a generous, honorable and honest man.

In the legacy of E. I. du Pont were genuine and important accomplishments that lived on long after the eulogies had faded from memory. Chief among them was the powder business. He had given his adopted nation a superior and much-needed product as well as a superbly organized system for producing it. Beyond that, he had helped to infuse an innovative, scientific spirit into American manufacturing. E. I.'s mill plan reflected French expertise and techniques but also embodied his own innovations. In 1804 he had patented a modification to an automatic sifter he had observed at the Essonnes works that improved the separation of caked powder into grains of varying sizes. The water-powered device did the work of six men and did it around the clock.[13] In 1822 the DuPont mills became the first in the United States to grind powder with large, rolling wheels rather than the less efficient and more dangerous up-and-down pounding of the old "stamping" method.

Like Benjamin Franklin and Thomas Jefferson, E. I. du Pont considered science a gentlemanly pursuit. Before leaving France in 1799, he made frequent visits to le Jardin des Plantes, the Parisian botanical gardens. E. I. even listed "botanist" as his occupation on his passport for the trip to America. He brought seeds and plants from France for his gardens at Eleutherian Mills and sent American seeds back to botanists in Paris. In 1830 he experimented with sodium nitrate from Chile, a possible substitute for the more expensive potassium nitrate from India, but concluded that the Chilean nitrate's tendency to absorb moisture made it useless as a black powder ingredient, at least at that time. Still, he filed away his research results for the future.

E. I. imported hardy Merino sheep, developed originally in Spain, for Victor's woolen mill across the river. He joined several groups, such as the Philadelphia Society for Promoting Agriculture, the Pennsylvania Horticultural Society, and the American Philosophical Society. E. I. du Pont's interests

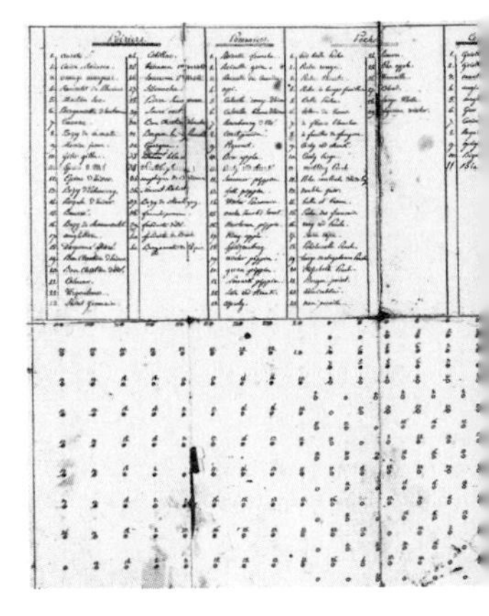

DuPont's first office, adjacent to the family residence, was constructed in 1837 by Alfred Victor du Pont.

Jacques Antoine Bidermann was one of the original investors in DuPont. His son, James Antoine, married E. I.'s daughter, Evelina, and headed the DuPont company for three years after E. I.'s death in 1834.

Margaretta La Motte was the wife of Alfred Victor du Pont. Alfred took over leadership of the company in 1837, continuing the pursuit of technological innovation.

Company records show that by the time E. I. died in 1837, DuPont mills had produced 627,161 kegs of powder.

in botany, chemistry and animal husbandry established a commitment to scientific discovery in all its diversity that has, over the years, woven itself into the very fabric of his firm.

The DuPont Company that E. I. left in 1834 was a substantial legacy. But opportunities are often squandered and businesses lost after a founder's vision and judgment fade with his memory. Upon E. I.'s sudden death, James Bidermann became director and senior partner. Assisted by Alfred as partner and Alexis as superintendent, Bidermann managed the business with quiet capability. He reorganized the firm as a family partnership among E. I.'s seven sons and daughters, retiring in 1837 when Alfred was ready to take over.

Conditions were favorable for the company's growth. Pioneers along the advancing frontier needed powder to clear tree stumps and shoot game. Settlement of the trans-Appalachian interior was creating new domestic markets for American farmers and manufacturers. Improvements in transportation were lowering distribution costs, putting a much broader market within reach.[14] The company's location near the Delaware River ensured service to international markets, but increasing quantities of powder were going to inland merchants by wagon and canal barge.

The construction of a national canal network during the 1820s and 1830s harnessed America's waterways to overcome the enormous problems that road construction posed in so vast a land. The excavation of one particularly rocky, three-mile stretch of the Erie Canal in 1823 consumed 14 tons of explosives, most of it from the DuPont mills. Railroads were still too dangerous to carry powder: Early steam engines showered sparks from their funnels. But the railroads' construction of embankments, cuts and grades consumed prodigious amounts of black powder. The railroads also made large coal shipments possible, and their steam engines consumed coal in great quantities. In 1859 DuPont purchased and refurbished a powder mill at Wapwallopen, Pennsylvania — its first plant outside the Brandywine — to provide blasting powder for the anthracite coal mining industry growing in the state's northeast corner.

Prompted by growing commercial activity throughout America and by government orders for powder during the Mexican-American War in the late 1840s, Alfred expanded the Hagley Mills still farther down the river into what became known as the Lower Yards. He also boosted employee benefits. In 1848 he extended the company's pension system to cover the widow of any worker killed in an accident unrelated to an explosion. He financed the transatlantic passage for workers' families — most of them from Ireland — and provided them with free medical services. Conscious of the dangers the du Pont family shared with the workers, Alfred also noted the special responsibilities of mill ownership. "We have a right," he said, "to expose ourselves without need if it is our pleasure to do so, but we are bound to protect the lives of our people to the best of our abilities."[15]

Alfred also shared his father's interest in research and technological innovation. He enjoyed working in the chemistry lab, exploring the various properties of powder. He evaluated a new explosive, guncotton, but concluded it was too unstable for large-scale manufacture. He also replaced the old water wheels with new, powerful turbines. When an 1835 coopers' strike cut off the supply of barrels, Alfred designed machinery that made barrel staves and eliminated the need for specialized skills. In 1837 he built DuPont's first office building next to the family's Eleutherian Mills residence, part of which had doubled as office space for 33 years.

But Alfred did not share his father's administrative skills or his careful attention to finances. Despite increased sales, the company was slowly sinking into debt. In January 1847, 10 years after Alfred had taken the company helm, his brothers and sisters — acting as his business partners — insisted that he get the company books in order. He had not produced so much as a balance sheet for the past year.

DuPont's First Office Building 1837
James Antoine Bidermann
Margaretta La Motte du Pont
DuPont Powder Works 1842
Alfred V. du Pont

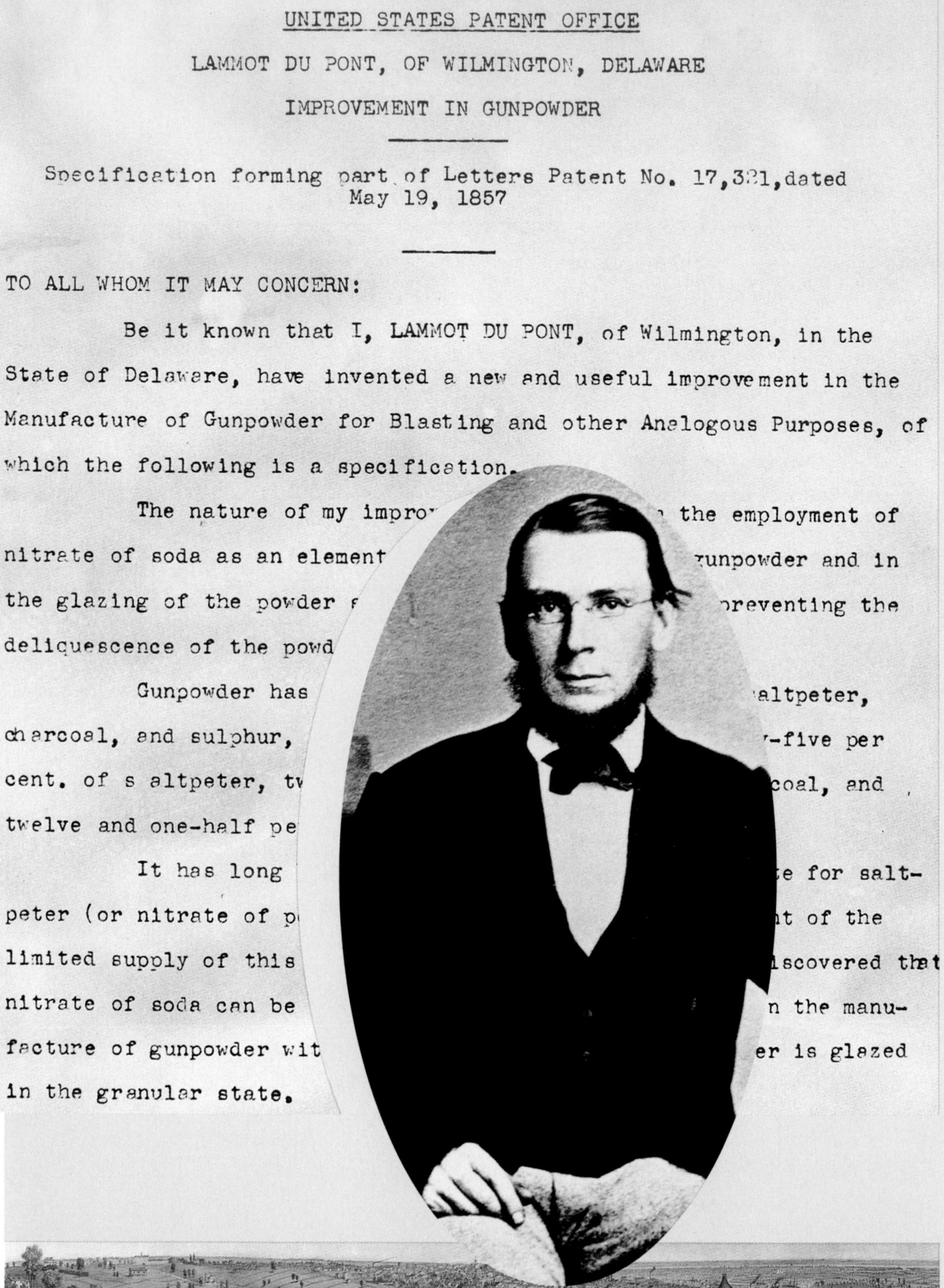

UNITED STATES PATENT OFFICE

LAMMOT DU PONT, OF WILMINGTON, DELAWARE

IMPROVEMENT IN GUNPOWDER

Specification forming part of Letters Patent No. 17,321, dated May 19, 1857

TO ALL WHOM IT MAY CONCERN:

Be it known that I, LAMMOT DU PONT, of Wilmington, in the State of Delaware, have invented a new and useful improvement in the Manufacture of Gunpowder for Blasting and other Analogous Purposes, of which the following is a specification.

The nature of my impro[...] the employment of nitrate of soda as an element [...] gunpowder and in the glazing of the powder [...] preventing the deliquescence of the powd[...]

Gunpowder has [...] saltpeter, charcoal, and sulphur, [...]-five per cent. of saltpeter, t[...] coal, and twelve and one-half pe[...]

It has long [...] e for salt-peter (or nitrate of p[...] t of the limited supply of this [...] iscovered that nitrate of soda can be [...] n the manu-facture of gunpowder wit[...] er is glazed in the granular state.

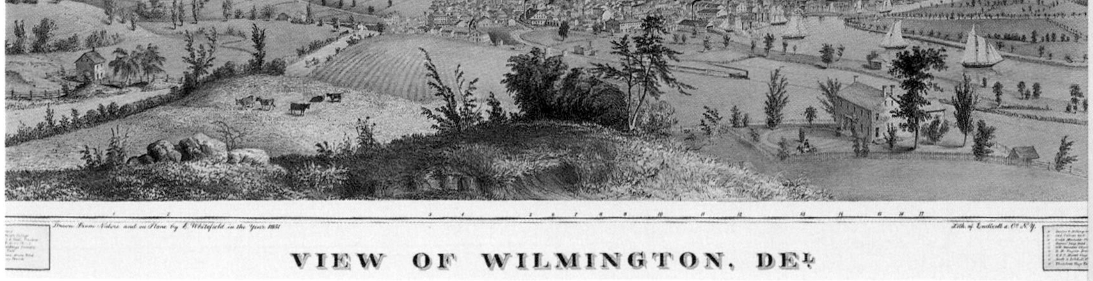

VIEW OF WILMINGTON, DEL

Just a few months later, on April 14, 1847, one of the powder mill buildings exploded, compounding Alfred's problems. Round-the-clock production schedules necessitated by the demands of the Mexican-American War contributed to escalating worker fatigue and carelessness. The first blast rained burning debris on neighboring structures, and shortly there were more explosions. To Ann du Pont, wife of Alfred's cousin Charles, the calamity seemed to transcend the Brandywine Valley. It seemed, she wrote, "not to be local but a crash of the world." Workers' families rushed downhill from their homes to the plant gate. "The shrieks of the wives and the children so soon made widows and orphans," Ann continued, "rose in sad succession to the preceding horror."[16] Eighteen workers died that day, adding nine widows to the six already on the DuPont pension plan. Yet the mills that were undamaged had to be kept going, filling orders while repairs were made and the company shored up the shattered morale of its workers.

By the time Alfred's sibling partners finally persuaded him to step down in 1850, the stress of the previous three years was evident. He had accomplished much and was reluctant to leave his post. The company had produced 2.5 million pounds of powder the previous year, and in 1850 *Scientific American* wrote glowingly of "the most extensive powder-mills in the world." But Alfred also was exhausted, careworn and ill. His tough-minded brother, Henry, took over a thriving, if disorganized and debt-ridden, business.[17]

Bolstered by English and French orders during the Crimean War and by the demands of the California Gold Rush, company sales increased between 1850 and 1855 at an average annual rate of 22 percent. Canal construction continued while railroad building boomed. By 1860, the year after DuPont opened its Wapwallopen mills, the Pennsylvania mines were producing 9 million tons of coal annually.

Henry would lead the firm for nearly 40 years, the longest tenure of any DuPont executive to head the company before or since. For more than 30 of those years he was assisted by Alfred's son, Lammot. Lammot finished his chemistry studies at the University of Pennsylvania in 1849 and went to work in the mills with his older brother, Eleuthère Irénée II, and his uncle, Alexis. In his father's last year as head of the firm, Lammot helped with various research projects, such as the testing of guncotton for the U.S. government. But the transition from college to the mills was not easy. In 1849, during his first four months on the job, 19-year-old Lammot spent long hours supervising operations, learning firsthand how fatigue and routine could wear down a worker's alertness to danger. He lost 30 pounds in the process.

On one occasion early in Lammot's career, the danger extended even beyond the plant's boundaries into the Borough of Wilmington. Three "powder schooners" were hauling DuPont's wares down Market Street on their way to wharves on the Delaware River. Each wagon carried nearly two tons of explosives in 25-pound barrels. They were required to keep a quarter-mile's distance apart for the same reason that the mill buildings were separate — to minimize damage to the whole in the event of an accident to a part. But on May 31, 1854, the worst possible thing happened in the worst possible place. For reasons never learned, the three powder wagons were traveling too closely together when one of them exploded, detonating the other two. The teamsters, their horses, and two citizens were killed, and buildings on Market Street sustained extensive damage.

The company's relations with the community suffered immediately and grievously. Alexis, the first family member to reach the scene, was met by an angry mob calling for the lynching of any du Pont who could be found. Fortunately, Lammot soon arrived, accompanied by some men from the mill. Together they faced down the mob and quelled its appetite for curbside justice. The company paid nearly $100,000 to restore damaged property, but only time could repair its damaged prestige in Wilmington. To help pacify its citizens and ensure their safety, DuPont built a road bypassing the city on which its product — "ticklish," as one worker put it — could be safely hauled to new, isolated wharves on the Christina River. To this day the bypass is called "DuPont Road." Even so, Wilmington passed a law prohibiting the transport of black powder within its limits.

Twenty-five years after its founder's death, the DuPont company had preserved and expanded E. I.'s vision of meeting a fundamental need with a superior product, manufactured by a skilled, cooperative work force using up-to-date technology. When Alfred retired in 1850, he and Lammot took up E. I.'s old interest in refining sodium nitrate. Lammot continued the research after Alfred's death and in 1857 patented a new method that rendered the substance suitable for blasting powder. The formula he developed allowed a higher percentage of oxygen and nitrogen, which burned off the nitrate impurities that made powder damp. This major advance in powder making reduced both the cost of the powder and the company's dependence on British-controlled potassium nitrate from India.

Henry sent Lammot abroad early in 1858, partly as a reward for his discovery and partly so that Lammot could survey powder-making operations in Britain and continental Europe. Lammot returned to the Brandywine three months later with drawings, copies of patents, laboratory equipment, and a cache of good will gathered during his visits to mills in France, Belgium, Germany, England and Ireland. The papers in his suitcase and the new ideas whirling around in his head were the stuff of innovation, an enthusiasm not always shared by his more conservative uncle Henry. The company's immediate future, however, would be determined most of all by larger, national events. In the West, controversy over Free Soil had already split "bleeding Kansas." As tensions between the North and South escalated and southern calls for secession grew louder, the nation itself came closer to being torn apart. Soon Lammot du Pont would return to England, not at the request of his uncle but at the behest of a Republican administration embroiled in civil war.

Lammot du Pont received a patent dated May 19, 1857, for his invention of soda nitrate. Explosive "B" blasting powder was the first notable change in black powder in 600 years.

In 1858, Lammot went to Europe aboard the ship *Arabia*. He befriended the captain of the ship and shared his fine Cuban cigars with him. One day Lammot also shared the fact that he had four pounds of black powder in his room to exchange for European samples. The alarmed captain, perhaps with an eye on Lammot's cigar, promptly ordered the black powder removed to the ship's magazine.

***BOTTOM:* By 1851, Wilmington was a city of some 15,000 residents. The economy was shifting from a community of merchants to manufacturing, such as shipbuilding and leather tanning.**

CHAPTER

2 FAMILY FIRM, GROWING NATION

RIGHT: In this Matthew Brady photo, Admiral Samuel Francis Du Pont stands before the Parrott Gun aboard his ship *Wabash.* One of the largest guns of its day, it fired DuPont Mammoth Powder.

BELOW RIGHT: Samuel Francis Du Pont (1803–1865) as a young officer. He was the son of E. I.'s brother Victor. Unlike his relatives, S. F. spelled his name with a capital "D."

OPPOSITE: Flag officer Du Pont captured Port Royal, South Carolina, for the Union November 7, 1861.

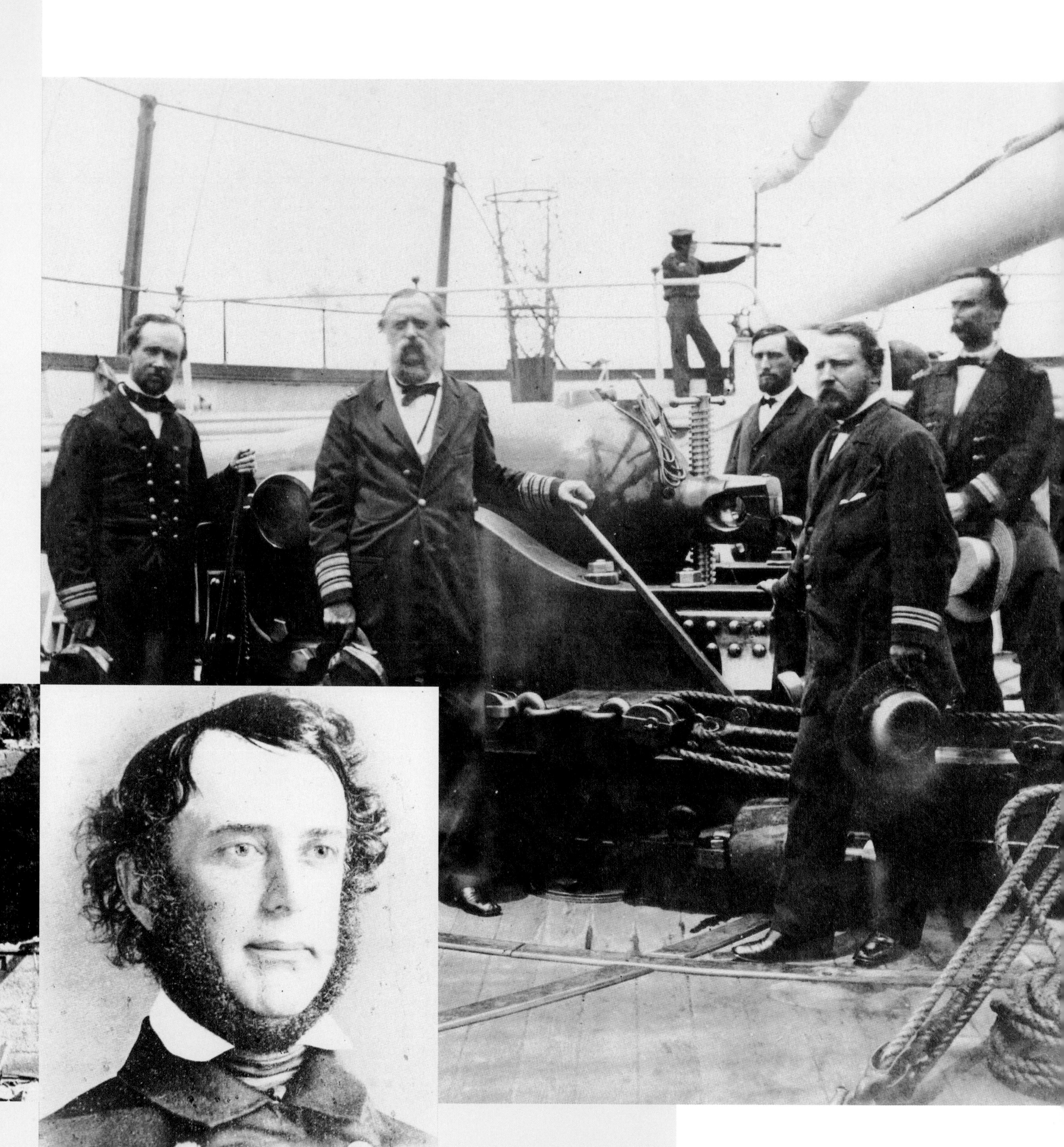

A tall, gray-eyed American stepped off the *Africa*'s gangplank and faded into the mid-November gray of London's dockyards. He was on a mission of misdirection, trying to make his activities seem routine and minimizing any connection with the U.S. government or the Civil War then raging in America. Lammot du Pont had come to England to buy up every last gram of saltpeter that $3 million of U.S. Secret Service money could buy, and whatever more his own credit could add. The Union depended on his success in a country that had close economic ties, and considerable political affinity, with the secessionist South.

Ironically, Lammot's uncle, Samuel Francis Du Pont, complicated an already delicate assignment. That same month, November 1861, Samuel Du Pont had boosted Union spirits with a daring naval assault on Port Royal, South Carolina. Flag officer Du Pont's capture of a key port in the heart of the Confederacy earned him promotion to Rear Admiral. Yet even as his ships' cannons overwhelmed the batteries of Fort Walker, Du Pont knew that the Union's gunpowder supplies were dwindling. He had sensed from the war's opening salvo at Fort Sumter back in April that the Southern "rebellion" would not be suppressed as easily as that term implied. His suspicions were confirmed as Union stockpiles of saltpeter fell to dangerously low levels. So Henry du Pont had been summoned to the White House on October 30, and within a week Lammot was on a steamer to England.

Ten days after his arrival, Lammot had arranged for four ships to begin loading the nearly 2,000 tons of saltpeter he already had acquired. But suddenly the loading stopped. Quiet cooperation from the British had turned into open hostility, and the Prime Minister, Lord Palmerston, called for a halt to the saltpeter sales. What had happened? Admiral Du Pont's blockade of Confederate ports had been a bit too successful. On November 9, Captain John Wilkes, commanding a ship in Samuel Du Pont's fleet, stopped the British mail ship *Trent* in Southern waters and seized two Confederate agents. The North cheered, but England fumed at this violation of her neutrality. Lammot, seemingly checkmated, returned to America for further orders while diplomats tried to calm the storm that Wilkes had stirred up. The United States feared pushing England into open support of the Confederacy and dreaded the loss of British saltpeter. So the American government soon apologized for Captain Wilkes' action, Britain confirmed its neutrality, and Lammot returned there early in January 1862 to complete his mission. Lammot brought back enough saltpeter to continue the war for another three years, which, as it turned out, was exactly what the defeat of the Confederacy required.

The four-ton, turbine-driven, iron rollers of the DuPont mills, rotating in pairs at nine revolutions per minute, ground out nearly 4 million pounds of gunpowder for the military during the Civil War, about 40 percent of all U.S. supplies. At the same time, DuPont managed to meet about half the needs of its commercial customers even though it increased its total number of workmen by only four.[1] Despite such economies, the costs of producing gunpowder rose steadily during the War, leading to the first cooperative meetings among the industry's major producers — DuPont, Hazard, Oriental and Smith & Rand. Lammot orchestrated this effort to balance a reasonable level of profit for the manufacturers with an affordable price to the government and to defeat congressional proposals to increase taxes on saltpeter and commercial powder.[2]

Lammot's leadership was due partly to DuPont's size and prestige, partly to his personal reputation as a skilled chemist and businessman, and partly to his confident, even jaunty, personality. Nevertheless, there was no question of who headed the firm back on the Brandywine. Henry du Pont disliked traveling, but at home he exercised leadership with relish. Just as he had taken up the slack reins of the company's management back in 1850, Henry now took firm control of Delaware's state militia, whose allegiance seemed uncertain. Appointed Major General in May 1861, 49-year-old "General" Henry purged the ranks of Confederate sympathizers and helped to ensure both the loyalty of the militia and the safety of the powder works from spies and saboteurs.

Henry personified the self-reliance so often associated with 19th century entrepreneurs. With a flaming red beard and a strong, burly physique that matched his forceful character, he defied any challenge to the status quo. Few men in the 19th century did more than Henry du Pont to change the face of America.

Henry was born on August 8, 1812, in the Eleutherian Mills house that his father, E. I. du Pont, had built. He graduated from the U.S. Military Academy at West Point in 1833 and served a year on the western frontier before resigning his commission to return to the Brandywine after E. I. died. "Coming home to the powder" was the family's way of putting it.[3] Henry was unshakably loyal to the firm and to the family, and he expected as much from every du Pont and every DuPont employee. No aspect of the company's operations was too minor for his close attention: he personally drew up every loading list for the powder schooners passing through the company gates.[4]

In the mid-1880s Henry's attention to detail narrowly averted a catastrophe. Returning home one night he heard a whistling noise coming up the hill from the powder works. He rousted foreman Thomas Kane from his house and the two hustled toward the noise. At the glazing mill their lantern lights revealed an overheated shaft, glowing red-hot just where it entered the mill from the turbine. Kane immediately shut down the turbine while Henry doffed his silk hat and handed it to his foreman. Using the sturdy hat as a makeshift bucket, Kane dipped water from the millrace and threw it onto the shaft. The hot metal steamed and sizzled but cooled rapidly enough to dampen the danger of an explosion.[5]

During Henry's long tenure of nearly 40 years as head of DuPont, his boots scuffed shallow, oblong tracks in the wooden floor beneath his spartan office chair. Nearby, beloved greyhounds stretched lazily in the heat of the stove. General Henry rode his carriage through the mills every day, with the dogs loping ahead to announce his inspection.[6]

The bark of Henry's demand for fealty was fierce and

LEFT: **Lammot du Pont kept Wilmington headquarters informed by telegraph during his journeys.**

BELOW: **Explosions were a constant danger at the DuPont powder works despite arduous safety precautions. The first major explosion occurred in 1815. Long-time employee Pierre Gentieu took this photo of a fire's aftermath on July 5, 1889.**

genuine, but his bite on loyal miscreants was softly paternal. If plant superintendents dismissed an employee for drunkenness — and if the man had a large family to support and was willing to pledge his best effort for future sobriety — Henry might find a place for the man repairing the stone walls on his extensive farm properties or performing some other task around the mills.[7] By the end of the Civil War, Henry was Delaware's wealthiest citizen, well able to afford the nearly 2,000 acres of farmland he purchased for his many experiments in agriculture and cattle breeding. Henry never stinted on time or energy in the office, but his greatest enjoyment was the life of a country squire.

The efforts and energy of Henry, Lammot and 218 workers and their families during the Civil War had made E. I. du Pont de Nemours and Company the nation's premier powder manufacturer. The United States had suffered greatly during the war: more than 600,000 lives lost and widespread physical destruction, especially in the South. But it also emerged from the conflict with an energized industrial capacity, an expanded rail transportation system and a new sense of the possibilities contained within thousands of miles of untapped natural resources. The triumphant Republican Party threw its support behind business and manufacturing interests, unleashing phenomenal expansion and change in American life. Between 1877 and 1893, railroad companies laid more than 100,000 miles of new track. Efficiencies brought about by standardized rail gauges and coordinated scheduling cut the costs of rail shipping by half, bringing cheaper goods to and from America's burgeoning cities.

Communications and transportation grew up together during and after the war, as copper telegraph wires were strung alongside the railroad tracks. The dots-and-dashes of Morse code telegraph messages were counterpoint to the clickety-clack rhythm of train wheels, and both created bright possibilities for coordination and efficiency in American industry. A telegraph line linked San Francisco with the East

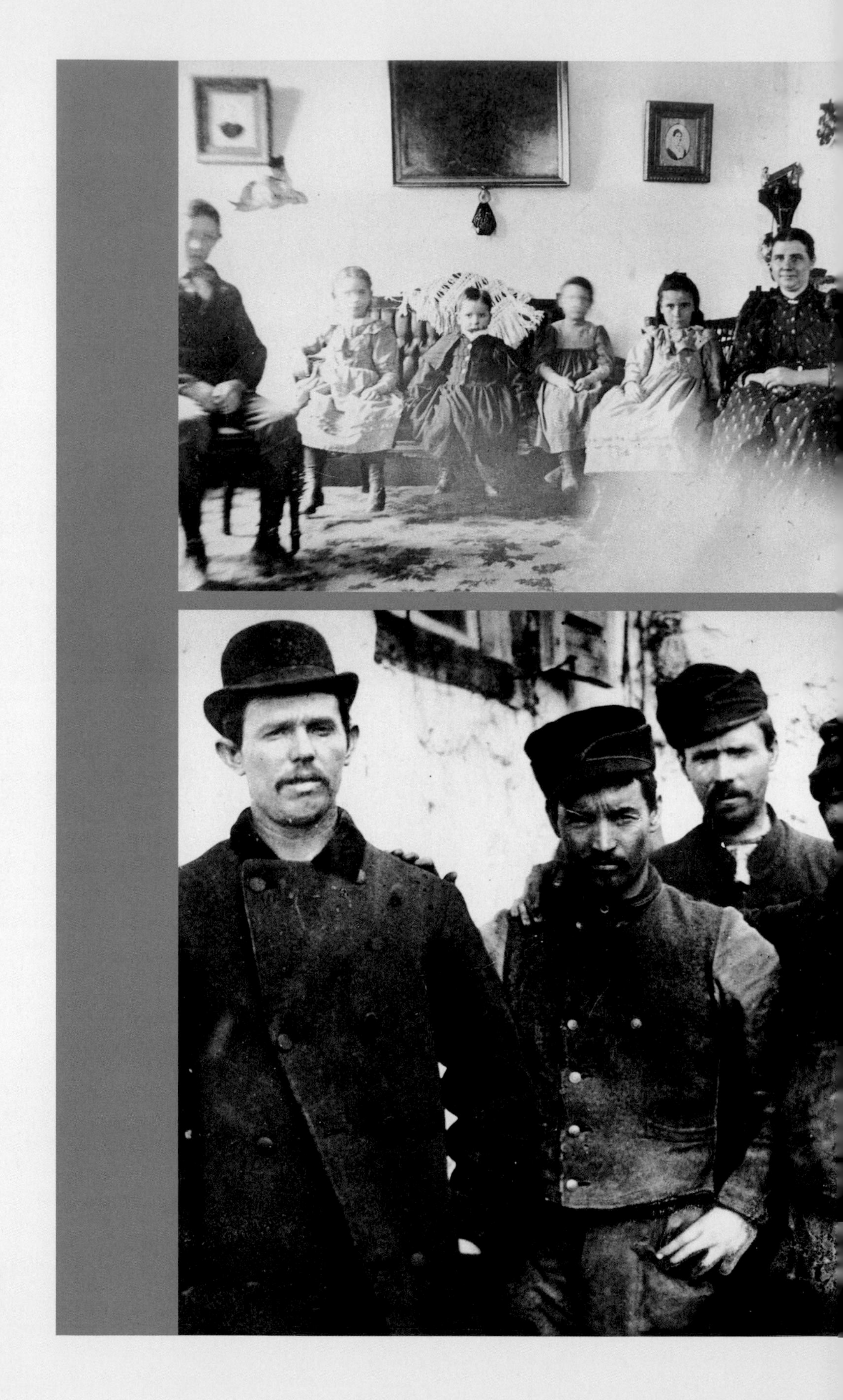

A CORE VALUE OF DUPONT SINCE ITS FOUNDING HAS ALWAYS BEEN FAIR AND RESPECTFUL TREATMENT OF PEOPLE. DUPONT WAS ONE OF THE FIRST AMERICAN COMPANIES TO PROVIDE A PHYSICIAN FOR EMPLOYEES, PAY OVERTIME AND NIGHT PAY, START AN EMPLOYEE SAVINGS PLAN, GRANT VACATIONS AND FUND A BONUS PLAN.

NOTABLE

> Where did the Confederacy get its powder? The South had only two small powder mills, in Tennessee and South Carolina. Northern mills like DuPont's supplied most of the southern states' needs before the war, and when war broke out, those supplies dried up. Confederate forces promptly seized hundreds of tons of powder held in federal arsenals and forts in the South, and in the magazines of companies like DuPont, Hazard, and Smith & Rand. Most shells and bullets fired at Northern troops in the early months of the war were propelled by Northern-made powder, although blockade-running ships also brought English powder into Southern ports. Neither side had expected a long war, and it took the Confederacy a full year to build a new powder plant at Augusta, Georgia. Still, throughout the war Southern supplies remained short and prices high. Over the course of the war, DuPont produced 4 million pounds of military powder. This exceeded the Augusta plant's production by 1 million pounds — a record in which DuPont took great pride.

> Samuel Francis Du Pont was the son of E. I.'s brother Victor. He began his naval training at the then customary age of 12. Rising steadily through the ranks, he patrolled the California coast during the Mexican-American War and served on the commission that established the Naval Academy at Annapolis. From the deck of his ship, the *Wabash*, he directed the successful attack on South Carolina's Port Royal in the early months of the Civil War. He was rewarded with promotion to the rank of Rear Admiral. Ordered against his better judgment to attack Charleston with the new, iron-clad "monitors" in April 1863, Admiral Du Pont was forced to retreat in the face of superior firepower. He died at age 61 on June 23, 1865, just a few months after the end of the Civil War. A monument dedicated to his memory was erected in Washington, D.C., in what is now called Du Pont Circle.

> Admiral Francis Samuel Du Pont was not the only family member to gain national recognition as a Civil War hero. Henry's son, Henry Algernon du Pont, was a West Point graduate who commanded an artillery battery during a number of battles in the Shenandoah Valley in the last two years of the war. On October 19, 1864, he rallied his troops in the face of a surprise dawn attack by Confederate General Jubal Early's soldiers at Cedar Creek, buying valuable time that allowed Union forces to regroup. Major du Pont was awarded the Medal of Honor for his bravery.

> Delaware was a border state, and during the Civil War its two slave-owning, agricultural counties leaned toward the Confederacy, while its northern county was mostly pro-Union. A sizable minority of citizens in the du Ponts' home county of New Castle were Southern sympathizers, but the DuPont leadership had always been staunchly pro-Union and anti-slavery, and no slave labor had ever been used in the company's operations. When Henry du Pont took charge of the state militia in 1861, he demanded that each man take an oath of allegiance to the United States. Governor Burton suspended General du Pont's order, whereupon Henry called for Federal troops to assert Union control of the militia. The troops and Henry's determination helped place Delaware firmly in the Union camp.

> In the 1890s, DuPont organized shooting tournaments as a means of introducing people to the superior quality of the company's smokeless powder. These shooting tournaments were organized into daily contests for both experts and amateurs. Cash purses for each totaled $1,220 — no small sum in those days. However, the strict tournament rules included the stipulation that contestants "must burn only DuPont smokeless powder." DuPont sold the ammunition at less than cost in order to provide sportsmen the chance to "be in full accord with the thousands who have already pronounced DuPont Smokeless 'the best Nitro in the world.'"

> Alfred Nobel's invention of dynamite in 1866 opened a new world of possibility for large-scale engineering and industrial projects. Three times more powerful than black powder, dynamite was useless for most military purposes (it damaged the gun barrels) but highly effective in mining and in the construction of dams, tunnels and bridges. DuPont's Repauno plant was the company's first move into this new product. Dynamite proved useful in some unconventional ways, too. The U.S. Navy used the new explosive to break up icebergs in shipping lanes; the lumber industry used dynamite to "top" tall trees. And during times of drought, some Texans exploded dynamite suspended from balloons in attempts to blast rain out of the clouds.

The United States of America

TO ALL TO WHOM THESE LETTERS PATENT SHALL COME:

Whereas

These are Therefore

In Testimony whereof

ABOVE: **A Matthew Brady photo of Sophie Madeline du Pont (1810–1888), youngest of E. I.'s three daughters and wife of her first cousin, Admiral Samuel Francis. She was a conscientious family diarist and correspondent.**

RIGHT: **As railroad and bridge construction and mining proliferated in the United States with the help of DuPont powder, Lammot du Pont kept striving to improve the company's powder-making capacity and move DuPont in new directions. He received this patent for a gunpowder press in 1865.**

FAR RIGHT: **Employee Jack Reed helped build this horizontal press for Hagley Yard in 1899.**

in 1861; eight years later the eastward and westward thrusts of the nation's first transcontinental railroad were joined at Promontory Point, Utah.

The herculean engineering feats of the latter 19th century — the dams and reservoirs, tunnels and railways, harbors and bridges, mines and oil wells — all started with a big bang, and General Henry was determined that each bang would come from DuPont powder. By the end of the Civil War, the family firm already was too large for Henry to manage on his own, however much he strove to reach his ideal of personal stewardship. Lammot had earned Henry's trust and admiration and was rewarded accordingly with a larger share of the company's responsibilities.

Lammot's patented "Mammoth" powder, developed with Army Captain Thomas Rodman in 1858, had improved the effectiveness of cannons during the Civil War and helped protect Union artillerists from dangerous breech explosions. In 1865 Lammot patented a new, hydraulic powder press. Besides being more efficient, the press helped reduce the chances of explosions. There had been 10 such accidents at the mills during the war, taking the lives of 41 men.[8] During the war Lammot toiled alongside mill workers, "killing himself in the refinery, working day and night," Sophie Madeline du Pont informed her husband, Admiral Samuel Du Pont. "Lammot's time is all engrossed with innumerable cares," she continued, "for he plans all new inventions, overlooks buildings, etc., etc., — indeed is the *life* of the business."[9]

Henry gave Lammot enough latitude that the younger man, called "Uncle Big Man" by the du Pont children, felt he had room to grow on acceptable terms. This working alliance between two du Pont generations lasted for 30 years. The wartime "gentlemen's agreements" about prices and taxes with Hazard, Oriental and Smith & Rand led to further meetings in the early 1870s to address new issues of competition and pricing in America's fast-growing economy. As DuPont expanded to meet the growing nation's postwar needs, Lammot traveled widely and often to coordinate the firm's various operations. He became so accustomed to sleeping on trains that he once wrote to his wife from a hotel room in California, "I turned in to bed only to kick all night, as I missed the motion of the cars."[10]

America after the Civil War was an extraordinary arena for competition, and Americans became keenly aware that success, power and personal character were closely linked.[11] While reformers and social critics debated the morality of "survival of the fittest," many American businessmen came to the practical realization that "unbridled" or "ruinous" competition, as they called it, could undermine their chances for success. They decided, sometimes grudgingly, that it could be in their interests to pool resources with competitors.

DuPont and other explosives producers set aside their differences to prevent the collapse of prices that was threatened by the government's plan to dump its surplus military powder on the open market. Henry proposed buying up the surplus in one swoop, but the government turned him down. Such a deal would be too controversial at a time when corruption in many wartime government contracts was being exposed. Instead, federal officials sold the powder at public auction, where it was purchased between 1865 and 1872 at wildly fluctuating prices by dealers large and small, reputable and shady. The war had kept the large powder firms occupied with military production, thus opening the commercial market to new, smaller competitors. The result was great unpredictability in supply and demand, with major producers unable to plan raw materials purchases, anticipate labor needs or establish efficient production schedules. The industry leaders had a choice: either fight to the finish, destroying many of their own number in the process, or band together to tame the upstarts and impose order. On DuPont's initiative, they chose the latter course.

On April 23, 1872, the heads of America's seven

agreement with the same effect as if it had signed this agreement as a party hereto, included under the name of the Laflin & Rand Powder Co.

IN WITNESS WHEREOF the parties hereto have hereunto set their hands and affixed their seals the day and year first above written.

The Hazard Powder Co.

Powder Mills

Powder Co.

leading powder companies gathered in DuPont's New York sales office at 70 Wall Street to establish the terms of their new cooperation.[12] Some participants were skeptical, fearing that competitive pressures would eventually undermine any agreement they could reach. Others, like Lammot, were more optimistic about the effort's viability. Both viewpoints found expression in the bylaws of the Gunpowder Trade Association (GTA) that the companies formed that April. Lammot was elected president of the GTA and served six consecutive one-year terms.

The GTA's rules laid out elaborate monitoring procedures and monetary penalties for violation of agreed-upon price and production levels. They also marked out proprietary sales territories for participants, thereby bringing order to the national explosives market. In 1875 the GTA negotiated an agreement with the large California Powder Works to establish a "neutral zone" of western states that both entities could serve. The GTA also invited nonparticipating powder companies east of the Rockies to join. Most did. Those that did not, and that attempted to undercut the GTA's prices, suffered the consequence of a concentrated price siege that the powerful GTA members invariably won. Such tactics would be considered illegal today, but in the 1870s and 1880s, the United States had not yet formulated a set of laws regulating the trade practices of its new big businesses, and with the government unable or unwilling to regulate trade, the producers took it upon themselves.

The nationwide depression that hit in 1873 tested the self-discipline of the GTA. Despite breaches of the agreement by nervous sales agents who cut secret deals with favored customers, the association held together. The new consolidation of national markets required a major adjustment by these formerly independent sales agents, who now had to coordinate their decisions more closely with corporate headquarters and follow policies established by executives who often were hundreds of miles away. The agents lost a degree of valued autonomy, though many of them also believed that they gained security and prestige through their association with powerful companies like DuPont.[13]

In the 1870s and 1880s DuPont's policy of buying up other GTA members strained the association's spirit of cooperation but did not violate the basic principle behind it — that cooperation existed not for its own sake but to avert destructive competition. DuPont's secret acquisition of the Hazard Company in 1876 gave it effective control of the GTA. Purchases of controlling interests in other companies, such as the California Powder Works, the Lake Superior Powder Company and the Sycamore Powder Mills in Nashville, gave DuPont a reputation not just for making quality powder but also for deploying its formidable size and resources forcefully in the national struggle for business survival. By the early 1880s Laflin & Rand was the only major competitor left. Lammot could rightly say, "The GTA is only another name for Du Pont and Co.... Whatever is Du Pont's interest will be done even if the GTA dissolves."[14]

Buying up the competition was not the only DuPont growth strategy in the latter half of the 19th century. Lammot and Henry also tried their hand at investing in railroads and coal mines in order to reduce production and distribution costs. DuPont already operated an extensive powder works at Wapwallopen in the Pennsylvania coal region, so it seemed but a natural step when the company organized the Mocanaqua Coal Company in 1866 and named Lammot its president. Coal looked like a good investment in the boom years immediately after the war, but the industry was prone to labor strife and at the mercy of highly variable shipping rates imposed by railroad companies, who were engaged in their own struggle for "survival of the fittest." The Mocanaqua Coal venture limped along for several years before DuPont unloaded it in 1881.

One of DuPont's more successful ventures beyond the powder mills involved a new product that was at once very

ROW ONE, LEFT: **Workers at Wapwallopen Powder Mills.**

ROW ONE, RIGHT: **An August 24, 1886, terms of the Gunpowder Trade Association agreement were revised regarding prices of various powder grades in different regions of the country.**

ROW TWO, LEFT and MIDDLE: **Workers at Wapwallopen Powder Mills.**

ROW TWO, RIGHT: **By the 1890s, Colorado was an important DuPont market because of its mining industry. This was the Denver office.**

ROW THREE, LEFT: **To expand beyond military markets, DuPont advertised its powder to miners.**

ROW THREE, MIDDLE: **DuPont began to invest in railroads and coal mines to lower production and transportation costs. The Wapwallopen works in Pennsylvania's coal region was the beginning of DuPont's national growth.**

ROW THREE, RIGHT: **This was DuPont's office on Wall Street, New York City.**

TOP LEFT: **Lammot du Pont built his Repauno dynamite plant in Gibbstown, New Jersey, on 750 acres of marshland. Several Lenape Indian burial mounds were discovered during construction, along with old coins and utensils of early English settlers. Lammot, always interested in Indian lore, added them to his artifacts collection.**

TOP RIGHT: **William du Pont (1855–1928) worked for DuPont for most of his career, leaving only to help Lammot in his dynamite venture at Repauno, New Jersey. William succeeded Lammot as president of this offshoot company, moving the headquarters to Wilmington.**

MIDDLE and BOTTOM: **E. I. du Pont II worked in the powder mills for almost three decades. He helped promote such advances as metal (rather than wooden) kegs for safety. The photo is a Daguerreotype.**

E. I. II and his wife, Charlotte, died in 1877, leaving five orphans at Swamp Hall on Breck's Lane, Wilmington. The children refused to be taken in by relatives, so, under the care of oldest sister Annie and several servants, they raised themselves.

promising and extremely dangerous: dynamite. Alfred Nobel's high explosive combined powerful but highly volatile nitroglycerine with a clay absorbent that stabilized it. Dynamite was not yet foolproof, but Lammot was confident that it could be improved with more research. He also was convinced that dynamite would be vital to DuPont's future. Already, "nitro" was three times more powerful per pound than black powder and was outselling powder nearly 3-to-1. Lammot's enthusiasm for the new product strained his working relationship with Henry, who looked at nitroglycerine's safety record and declared his steadfast opposition.

The strain between the two men involved more than disagreement over dynamite. Lammot had been chafing under Uncle Henry's harness for a long time. His irritation surfaced in 1877 when his brother Eleuthère Irénée II and Irénée's wife, Charlotte Henderson du Pont, both died of illnesses, leaving five orphans in their Brandywine family home, Swamp Hall. Forty-six-year-old Lammot took the occasion to look into the company books and review Uncle Henry's management of the firm.

Lammot knew that the DuPont company, like Henry himself, dominated competitors by its strength and its commitment to solid, traditional standards. But Lammot also deduced from 1870 United States industrial census data that DuPont was the least efficient of all major powder companies. Over the years, innovations had been patched and grafted onto old structures. Workers often had to follow roundabout routes to carry materials and products from one stage in the manufacturing process to another, taking extra steps to accommodate the existing system. Steam power helped liberate powder production from the millrace, and Lammot persuaded Henry to add seven more steam engines to the mills' existing three between 1870 and 1874. Still, major improvement was difficult in an obsolete plant dependent on immovably thick walls and a fixed water supply.

Lammot also was critical of the firm's management. In December 1877 he proposed to Henry several changes that

would spread authority more widely among the six partners — Henry, Lammot, cousins Eugene and Frank, and Henry's sons, Henry Algernon and William. He also suggested higher salaries for the junior partners. Henry never rejected Lammot's proposals. Instead, he let the months pass by as if there were all the time in the world to discuss such things. In April 1878 Lammot resigned in frustration, though he stayed on for several more months to tie up the loose ends of his many DuPont duties.

In 1880, at Thompson's Point, New Jersey, across the Delaware River from Wilmington, Lammot realized a dream that had been pent-up during his years on the Brandywine. He began construction of a new dynamite plant that he named the Repauno Chemical Company. Lammot had probably been considering this move for several years. DuPont's 1876 purchase of a controlling interest in the California Powder Works had included a Cleveland subsidiary, the Hercules Company, that produced dynamite. In visiting Hercules, Lammot had learned much about bulk dynamite manufacture; in fact, he purchased Hercules a year after he opened the Repauno plant in May 1880.

Lammot had raised capital for Repauno from the "big three" powder companies, DuPont, Hazard and Laflin & Rand. Because DuPont owned Hazard, the company actually had a two-thirds share in Lammot's enterprise. Lammot remained a partner in DuPont and held a one-sixth share in Repauno, which he enlarged to one-third in 1882 by exchanging his interest in DuPont for a third of DuPont's Repauno stock. His future was now bound up in Repauno and dynamite, both of which looked very promising. Repauno's 200 workers produced 3 million pounds of the new explosive in 1881, the first full year of operations. But there were problems with large-scale production of dynamite that Lammot had yet to solve.

Nitroglycerine production left toxic acid byproducts that leaked from Repauno into the small streams and eddies

TOP: **Chemist Francis Gurney "Frank" du Pont (1850–1904) was a partner in the family company. He helped build the Mooar, Iowa, mills and assisted in the invention of smokeless powder for shotguns in 1893.**

BOTTOM: **West Point graduate Henry Algernon du Pont (1838–1926) distinguished himself in Civil War service and was awarded the Medal of Honor for heroism. He later served as a DuPont vice president but left the company for politics and other interests. He served in the U.S. Senate from 1906 to 1917.**

Smokeless Powder Seal

emptying into the Delaware River. Fishermen complained of the damage to sturgeon and shad and brought suit against Repauno. The company settled, but Lammot began experiments to capture and recycle the acids. On Saturday morning, March 29, 1884, he was in the laboratory when a workman rushed in to tell him that 2,000 pounds of nitroglycerin that was supposed to be cooling in a lead-lined tank nearby was beginning to fume. Lammot hurried to the scene to empty the mixture into a water tank but realized he was too late. He raced out of the building with four other men, all making the most of the few seconds before the explosion. Lammot's lanky frame lunged through the air for the safety of an earthen mound just as the impact of the exploding nitro smashed into him. All five men were killed instantly.

Frank and Eugene ferried across the Delaware as soon as they could but there was little to do except observe the deadly effects of the product their uncle Henry had kept at arm's length — with good reason, it now seemed. Solomon Turck, president of Laflin & Rand and a friend of Lammot's, took over at Repauno for two years until being replaced by Henry's son, William. The DuPont Company bought back Lammot's one-third interest in Repauno, becoming once again its majority stockholder. Letting business acumen override personal reservations, Henry honored Lammot's commitment to the new explosive, and Repauno soon became the country's largest dynamite producer. But Henry also made another massive commitment, this time to his old product, black powder.

In 1888, 76-year-old Henry sent Frank to the midwestern mining region near Keokuk, Iowa, to build a state-of-the-art black powder mill. In one sense, the new Mooar mills were a colossal monument to the past, for clearly dynamite, rather than black powder, was the up-and-coming product. In another sense, the Iowa mills were a forward-looking investment because black powder remained the coal miners' explosive of choice for many more years. Dynamite pulverized coal, blasting it into useless dust, whereas black powder simply

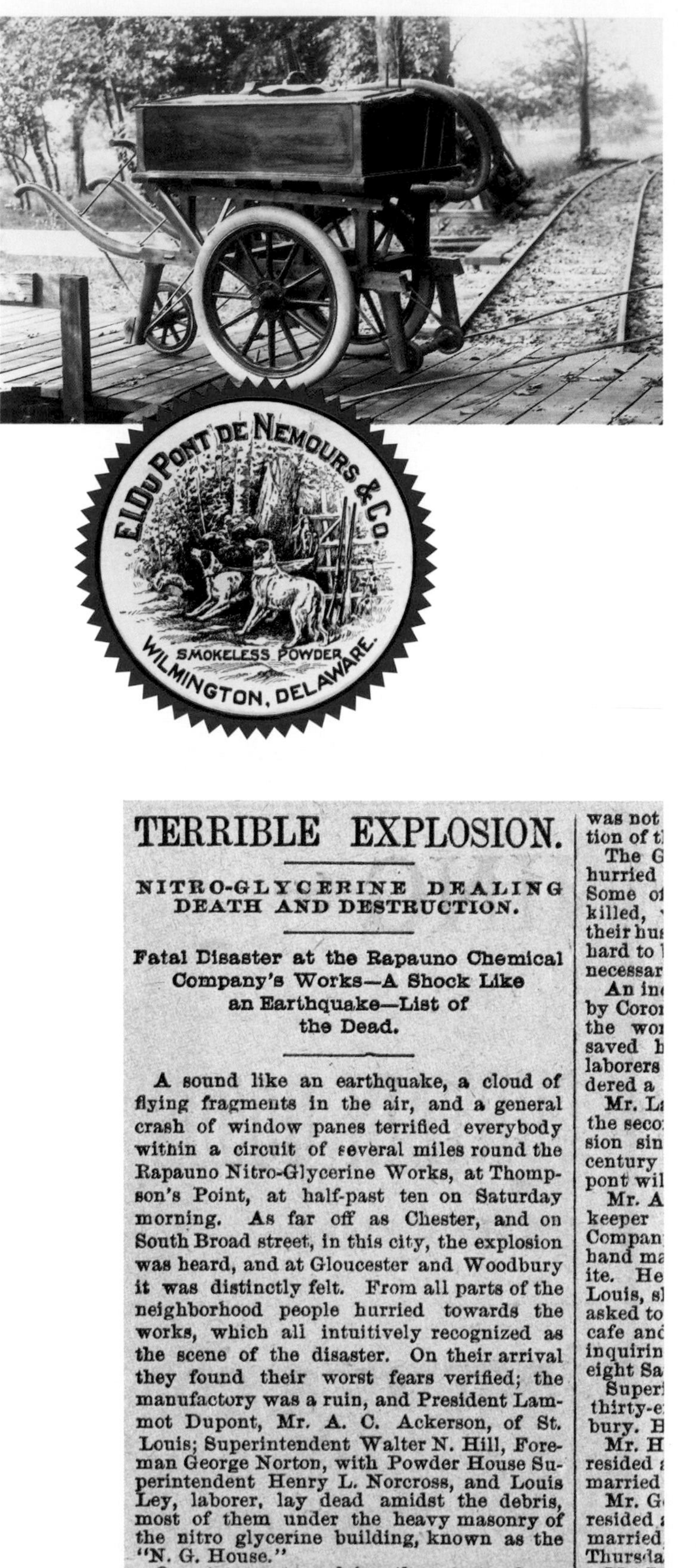

TERRIBLE EXPLOSION.

NITRO-GLYCERINE DEALING DEATH AND DESTRUCTION.

Fatal Disaster at the Rapauno Chemical Company's Works—A Shock Like an Earthquake—List of the Dead.

A sound like an earthquake, a cloud of flying fragments in the air, and a general crash of window panes terrified everybody within a circuit of several miles round the Rapauno Nitro-Glycerine Works, at Thompson's Point, at half-past ten on Saturday morning. As far off as Chester, and on South Broad street, in this city, the explosion was heard, and at Gloucester and Woodbury it was distinctly felt. From all parts of the neighborhood people hurried towards the works, which all intuitively recognized as the scene of the disaster. On their arrival they found their worst fears verified; the manufactory was a ruin, and President Lammot Dupont, Mr. A. C. Ackerson, of St. Louis; Superintendent Walter N. Hill, Foreman George Norton, with Powder House Superintendent Henry L. Norcross, and Louis Ley, laborer, lay dead amidst the debris, most of them under the heavy masonry of the nitro glycerine building, known as the "N. G. House."

One statement explains the catastrophe as the result of an attempt to save a large part of the nitric acid which in the process of manufacture had previously run to waste into a tank in the cellar, after the distillation of the nitro-glycerine. Just before the explosion

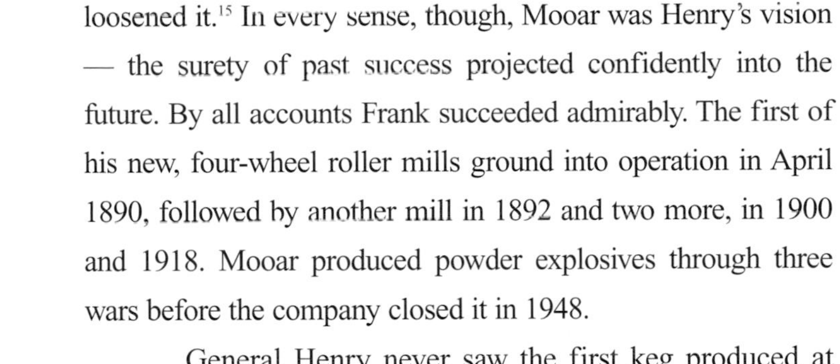

loosened it.[15] In every sense, though, Mooar was Henry's vision — the surety of past success projected confidently into the future. By all accounts Frank succeeded admirably. The first of his new, four-wheel roller mills ground into operation in April 1890, followed by another mill in 1892 and two more, in 1900 and 1918. Mooar produced powder explosives through three wars before the company closed it in 1948.

General Henry never saw the first keg produced at Mooar, nor did he see Frank's handiwork later become the largest black powder works in the world. He died quietly at Eleutherian Mills on August 8, 1889, his 77th birthday. Henry had been a remarkable man. Poised between the older values of rural America and the whirlwind of change that innovation, including his own, had helped bring about, he had steered a straight course through setbacks and successes.

Henry appears to have believed that the straight course — and the right course — would be as clear to anyone else as it was to him, and he provided little guidance to his immediate subordinates. Henry also left no instructions about selecting a successor, perhaps because he believed that this, too, would be so clear that there would be nothing to argue about. But his sons, Henry A. and William, did argue about succeeding their father, so the partners turned instead to Henry's 49-year-old nephew, Eugene, to head the firm. Eugene's younger brother Frank became plant superintendent. Eugene and Frank were skilled chemists, and they made important innovations at the mills.

By the late 19th century it had become commonly known that when cotton or some other cellulose material was soaked in nitric acid, it became highly explosive but left no smoke trail to obscure vision or give away a location. This substance was called "guncotton," or "smokeless powder." In 1893 Frank, assisted by Lammot's oldest son, Pierre, patented a smokeless gunpowder

OPPOSITE TOP: **A nitroglycerine barrow at the Repauno Chemical Co., Repauno, New Jersey, established by Lammot over the objections of his uncle Henry.**

OPPOSITE BOTTOM: **As if to validate his uncle's trepidations, Lammot died in an explosion at his dynamite factory in 1884.**

TOP: **This is the family of Lammot du Pont at St. Amour about 1894, some 10 years after his death. Back (l to r): Irénée, Alexis I., William K., Mary Aletta (later Laird), Lammot, Margaretta "Peg" (later Carpenter) Front: Isabella (later Sharp), Mrs. Mary Belin du Pont, Pierre Samuel (of Longwood), and Louisa (later Copeland).**

BUILDINGS: **Henry, intent on serving expanding railroad construction and mining markets in the West, in 1888 launched his most ambitious project to date, the construction of the colossal Mooar Mills near Keokuk, Iowa. The mills eventually became the largest black powder plant in the world.**

This was life along the Brandywine River near the end of the 19th century. The powder mills were built as a complete village, with houses, schools and churches all within close proximity. Children could deliver their fathers' lunches in pails directly to the mills. Alfred I. du Pont enjoyed music so much that he formed a band called the Tankopanican Musical Club, which played in the Wilmington area.

Stragglers
Time Book
May 1849
Sept 1855

for ordinary shotgun use. Eight years later DuPont opened a smokeless powder plant for military artillery at Carney's Point, New Jersey. Frank's work at the Carney's Point plant expanded DuPont's explosives products and gave the company valuable experience with nitrocellulose chemistry. But personnel and management practices followed by Frank and Eugene too often alienated DuPont workers, a situation that worsened after Henry died.

Eugene and Frank were bent on introducing what they regarded as modern methods of discipline and efficiency, and they cracked down on workplace practices formerly overlooked by Henry. "They weren't going to put up with that nonsense," recalled the daughter of the company bookkeeper.[16] In addition to enforcing discipline, they introduced policies that now seem like reasonable economies, though they were fiercely resented by some DuPont workers. Instead of hiring company employees to perform various masonry and carpentry work on the du Pont properties, for example, Frank and Eugene contracted outside for cheaper labor. Regular employees, who had come to think of extra income from this work as an entitlement, were baffled and outraged. A small number formed a secret group called the "Neversweats" that harkened back to rural peasant societies in Ireland that traditionally used violence to redress their grievances against landowning aristocrats.[17]

On the day after Christmas, 1889, the Neversweats burned down Frank du Pont's barn. The fire nearly spread to his home just a few yards away. A month later a company barn mysteriously burned down, killing Frank's favorite horse. Several months then passed uneventfully before an explosion of suspicious origin on October 7, 1890, killed 12 workmen, a mother and her baby, and caused $1 million in damage to company property. The du Ponts hired Pinkerton detectives to solve the mystery. Although their efforts led to the arrest and conviction of five people, barn burnings continued to occur, even as late as 1904. Undercover Pinkerton agents circulated regularly among DuPont's 300 employees along the Brandywine during this time, but no additional arrests were ever made.

Hoping to curb employee resentments, Frank took a cue from his nephew, F. G. Thomas, superintendent of the Mooar plant. In 1891 he established a workers' club in the abandoned Eleutherian Mills, E. I.'s original family home, built in 1803. Equipped with pool tables, showers, electric lighting, a bowling alley, a dance hall and non-alcoholic refreshments, the Brandywine Club attracted a few workers from Henry Clay Village, the employee housing area a mile or so away, although most thought it too long a walk after a hard day. DuPont followed the six-day work week that was typical for industries of the time — five 10-hour days, followed by nine hours on Saturday. In the late 1880s streetcar service made Wilmington's attractions more accessible from the Brandywine, but the long work hours and the high streetcar fares kept excursions to town rare.

There was no need to leave the village for food, clothing or other supplies. Four merchants competed for the general store trade, while John Wood and Son advertised "Drugs, Medicines, Chemicals, Toilet and Fancy Articles, Brushes, Soaps, Perfumery, and Patent Medicines." Most families grew their own vegetables. Dress- and shirt-makers visited homes to make clothes for those who didn't make their own. "Living conditions on Rising Sun Lane were the same as over on Breck's Lane," said William Buchanan, who grew up in the yards. "There was seven of us children in the family and with my mother and father that made nine.... We all lived up there and we got along fine." There was only one constable in the village, paid by the company, who never arrested anyone.[18]

Probably no du Pont ever felt so much at home among the workers in the village in the 1890s as Alfred Irénée du Pont, Lammot's nephew and E. I.'s great-grandson. DuPont employees remembered his boyish escapades as a member of the "Down-The-Creek" gang, which battled with the "Up-The-Creek" gang, and they appreciated his relaxed, rough-

TOP LEFT: **There were about five mule teams and one horse team delivering powder from the DuPont yards. All of the teams were discontinued in 1889.**

TOP MIDDLE and RIGHT: **Named for the prominent Whig politician, Henry Clay Village along the banks of the Brandywine was home to generations of DuPont workers.**

BOTTOM LEFT and RIGHT: **After Frank and Pierre's invention of smokeless gunpowder for shotguns, DuPont opened a plant at Carney's Point, New Jersey, to manufacture the new product. The site — across the Delaware River from Wilmington — was chosen because the Brandywine River complex was getting too crowded with mills and it provided better shipping facilities.**

TOP: **Troops guarded Carney's Point Powder Works during the Spanish American War in 1898.**

BOTTOM: **Famous for his knickers and socks, Alfred I. du Pont (1864–1935) often was called "Knickers." Even well after retiring, Alfred often organized excursions for the men who worked for him at the powder yards.**

FAR RIGHT: **T. Coleman du Pont (1863–1930) was one of three cousins to step up to buy the company in 1902 when it was about to be sold to its chief competitor. He did not remain with the company long, preferring other interests such as New York City real estate.**

and-tumble way with workers.[19] "Alfred I. du Pont was a friend of everybody's," recalled William Buchanan, whose father, Albert, was one of Alfred's closest boyhood friends. "He was just the same as any other powderman, dirty from the powder."[20]

Some called Alfred "Knickers" or "Short Pants" after his distinctive dress, but all respected his fighting spirit. Only 12 years old when his parents, Eleuthère Irénée II and Charlotte Henderson du Pont, both died in 1877, Alfred was one of the five orphans left in their Swamp Hall home on Breck's Lane, about 100 yards from the powdermen's houses. When Henry and the family elders determined to parcel them out to various relatives, the children armed themselves with a rolling pin, an axe, a pistol, a bow and arrow, and a shotgun and defiantly stood their ground. Alfred hefted the shotgun. Impressed by the children's show of unity and grit, Henry relented and let them stay in their own home. They rewarded his gamble by managing themselves creditably before heading off, one by one, for school.

Alfred joined his Kentucky cousin, T. Coleman, at Massachusetts Institute of Technology in 1882. Both enjoyed Boston nightlife, and Alfred befriended legendary bare-knuckles champion John L. Sullivan, with whom he sharpened his boxing skills. Though Alfred was no scholar, he was bright and a quick study on topics that interested him. He was fascinated by electricity and personally wired Swamp Hall in 1884 before extending the lighting further throughout the mills.

When Lammot was killed in 1884, Alfred decided to leave college and return to the powder mills. In October Henry hired him as an apprentice powderman. Alfred performed well, came into his inheritance when he turned 21 in 1885, and married Bessie Gardner, the beautiful and well-educated daughter of a Yale professor. In 1889 Henry sent him overseas to evaluate French progress in manufacturing brown prismatic and smokeless powder, both quick-burning propellants used by the military. When Alfred and Bessie returned, Henry was dead. During the partners' difficult discussions over Henry's successor, Alfred broke with the family tradition of showing deference to elders. Rather than wait politely for an invitation, he fought vigorously and successfully to be included as a junior partner. As he had already proven, he was not one to stand quietly by if he thought he was being shortchanged. He was also not one to shrink from a fight.

Soon after the American battleship *Maine* exploded in Havana harbor in 1898, the United States declared war against Spain and 34-year-old Alfred joined the Army. When ordnance officers familiar with his powder-making skills tried to persuade Alfred to stay on at the mills, Alfred agreed on the condition that Eugene and Frank would promise him complete control of production during the war effort. Alfred would brook no intrusions into his realm. He understood as well as his cousins the importance of technical proficiency, but he also knew how to build and sustain workers' pride. War production goals would never be met without both these ingredients.

Eugene and Frank agreed to Alfred's demand, though they wondered how he would ever manage to squeeze 20,000 pounds of brown prismatic powder *per day* out of the old plant. Working 18 hours a day, Alfred and the workers inspired by his example soon reached the 20,000 pound goal set by the ordnance officers — with an additional 5,000 pounds thrown in for good measure. These long hours proved to be Alfred's finest. They established beyond question his credentials as a true DuPont powderman, and they endowed him with the authority that only extraordinary success can give.

Despite the efficiencies he introduced as head of the firm, Eugene could not single-handedly manage the increasingly complex affairs of the DuPont Company in the late 19th century. Henry A. du Pont could see how entangled Eugene had become and suggested incorporation as a means of redistributing executive responsibilities. After nearly a century, the old partnership was dissolved, and E. I. du Pont de Nemours &

SOFT
DRINKS
SEGARS
TOBACCO
Eng by E. G. Williams & Bro NY

LEFT: Eugene du Pont (1840–1902) took over the presidency of the company after the death of his uncle Henry in 1889.

BELOW: Frank du Pont (1850–1904), Eugene's brother, became plant superintendent when his uncle Henry died. A skilled chemist, he and Lammot's son Pierre patented a smokeless gunpowder for ordinary shotgun use a few years later. This invention helped move DuPont into the modern explosives industry.

Company was incorporated in Delaware on October 23, 1899. Eugene became president. Frank and Henry A. became vice presidents, as did Alexis I. du Pont, Frank and Eugene's younger brother. Charles I. du Pont, great-grandson of E. I.'s brother, Victor, was the new company's secretary and treasurer, and Alfred was given the vague title of "director" and no executive responsibilities.

Old habits died hard, and incorporation did little to change the way business was conducted at DuPont. There was no new infusion of executive talent, no real reorganization or redistribution of duties, and the six officers retained all the company stock. Nominally a corporation, the firm still operated as a partnership with Eugene as sole leader. Young Pierre du Pont, Lammot's son, grew frustrated with the inertia among the company's elders and left his job at Carney's Point to pursue a streetcar venture in the Midwest. Alfred toiled on in the old mills, no more pleased than Pierre with the firm's leadership but happy enough to be producing powder, and satisfied that at least he had acquired a 10 percent interest in the company.

On January 28, 1902, Eugene died unexpectedly of pneumonia at age 61. Again, there were no plans for a successor and, more ominously, no volunteers when the officers met to fill the vacated presidency. To everyone's surprise, Frank declined the presidency. He had not yet told the other officers that he was very ill, and indeed he died just two years later. Nor would Henry A. take Eugene's place, for he had become involved in politics and was considering a bid for the U.S. Senate. Alexis I. was also in poor health and forthrightly said so. He too would be dead in two years. That left the two younger officers, Charles and Alfred. Remarkably, Charles was ill, as well, and died within a year.

Alfred had failed to attend the meeting to elect a new president, thinking perhaps that Frank's election was a sure thing. There was no great warmth between the two men. Alfred's absence was duly noted and gave the others an opportunity to discuss his suitability to be head of the firm, and no one objected when Frank ruled him out. Only one course remained, it seemed. The company would have to be sold to its friendly competitor, Laflin & Rand. The meeting adjourned on this gloomy note, and Frank went to tell Alfred the news. E. I. du Pont de Nemours & Company was about to be orphaned.

Alfred took the news with apparent calm that belied his intentions. He had fought the family before and won. But to win on an issue like this, he would need more than shotguns and rolling pins. Alfred rushed to his former MIT compatriot, cousin T. Coleman, whose financial acumen already had earned him a reputation as a shrewd businessman. T. Coleman agreed to help. On February 14, the DuPont company officers assembled in a meeting room at company headquarters on the Brandywine to decide on a selling price for Laflin & Rand. Alfred, ever the defiant boy from Swamp Hall, showed up in his grimy knickers and feigned sleepiness while Frank read the motion to sell the company. When he finished, Frank looked around the room for the sad but expected approval. Just before it came, Alfred calmly raised his hand to suggest a harmless amendment. Instead of specifying Laflin & Rand, could the motion simply read "to the highest bidder?" The amended motion carried without so much as a raised eyebrow. Then, as Frank moved to adjourn, the powder-flecked arm rose once more. Alfred stood and, with all the quiet pride he could muster, announced that *he* was prepared to purchase the company.

Frank exploded in opposition and Alfred's sang-froid disappeared in a verbal fury. The company was his birthright, his heritage, he protested. He was the oldest son of the oldest son of the oldest son of the founder. Could the officers not give him just one week to work out the finances and present them an offer? Henry A. finally broke the tension in the room. He thought that Alfred's proposal was sound, even right, and deserved a chance. At last the others agreed. Twenty-five years earlier General Henry had given the Swamp Hall orphans a chance to make it on their own. Now Henry's son was giving the same chance to Alfred to keep the firm in the family.

CHAPTER 3 THE “BIG COMPANY”

Henry Algernon du Pont followed the victorious Alfred out of the room. He was stirred by Alfred's appeal to family tradition and the honor of the du Pont name, but he also knew of Alfred's reputation for impulsive decisions. Henry chose his words carefully. "I assume, of course," he said, now that both were well out of earshot of the others, "that Thomas Coleman and Pierre are, or will be, associated with you in the proposed purchase?" Alfred assured him that they were, though he still needed final confirmation from Pierre. Henry gave his conditional support. Alfred climbed into his automobile, still a novelty on Wilmington's streets, and drove straight for Coleman's house on Delaware Avenue.

Alfred had called on Coleman several days earlier to discuss the purchase, and Coleman had been thinking hard since. His wife and cousin, Alice "Elsie" du Pont, who had grown up on the Brandywine, offered brief but sage counsel: "You know what it is like to be in business with your relatives." Coleman's father, Bidermann du Pont, and his uncle Fred (Alfred Victor) had left the Delaware branch of the du Ponts behind and struck out for Kentucky before the Civil War to find their fortunes, not in black powder but in paper mills and coal mines. Coleman had worked in those businesses, and the freedom he enjoyed there contrasted with stories he had heard about the way General Henry, Eugene and Frank handled things back on the Brandywine. Coleman was not sentimental about "coming home to the powder," but he had fond feelings for his Delaware cousins. Still, Elsie's words echoed in his mind. Did he know what it was like to be in business with relatives? Well, he concluded, relatives need not be more difficult than anyone else. Coleman was used to winning, and he was sure he was up to this challenge if he had the right plan.

Thomas Coleman du Pont was born to Bidermann and Ellen Coleman on December 11, 1863, when the winds of the Civil War were shifting in the Union's favor. He grew to a sturdy 6 feet 3 inches and 210 pounds, superbly athletic and bursting with confidence. Coleman studied mining engineering at the Massachusetts Institute of Technology, where he roomed with his cousin Alfred. He and Alfred enjoyed Boston's theaters more than their studies, and both left without finishing their degrees. Coleman returned to Kentucky to learn his father's coal mining business in the company town of Central City. Like Alfred, Coleman enjoyed the hearty comradeship of the working men and carried a rough-and-ready spirit into his supervisory and executive positions. For Coleman, high finance was not much different from a tug-of-war or a football match: the fellow who buried his doubts and weaknesses and stepped out with gusto and a grin was the one most likely to win, especially if he kept moving.

After eight years with the Central Coal & Iron Company, Coleman moved to the Johnson Steel Company in Johnstown, Pennsylvania, which specialized in making rails for the new electric streetcars in many American cities. Uncle Fred had invested heavily in his protégé Tom Johnson's steel company, and when Fred died in 1893, he left Coleman and Coleman's cousin, Pierre, substantial amounts of Johnson Company stock. Johnson's interests shifted to politics in the

TOP LEFT: Alfred I. du Pont became production chief of the company. In 1902, he blocked the sale of the firm to competitors while he sought his cousins' aid.

BACKGROUND: Many in the du Pont family had a fascination with cars and were among the earliest automobile owners in Delaware. This is Alfred I. in his 1902 single-cylinder Didion Bouton, the second car in Delaware.

OPPOSITE: Perhaps it was this experience with a muddy road that prompted Coleman to announce in 1908 that he was "going to build a monument a hundred miles high and lay it on the ground" so Delaware would have a concrete highway running its entire length. He did just that, creating what is now U.S. Highway 13.

BOTTOM: Coleman du Pont was a free spirit who could never stay tied to one project for very long.

OPPOSITE TOP: **Arthur Moxham worked with Coleman du Pont at the Johnson Steel Co. in Johnstown, Pennsylvania. He joined DuPont as director of the Development Department in 1903. Moxham encouraged researchers to experiment in chemical manufacturing to push DuPont in new directions.**

CIRCLES: **Pierre Samuel du Pont, unlike cousins Coleman and Alfred I., completed his studies at MIT. His experience as a chemistry student probably influenced the direction of DuPont, since he built up its experimental branch after World War I. When Pierre was a student, applied science was only beginning to become a respected field, but DuPont would help change that.**

early 1890s, and he went on to national fame as a congressman and reform mayor of Cleveland. When he entered politics, he picked Coleman to be the firm's general manager under president Arthur Moxham, an engineer who had helped Johnson adapt railroad tracks to streetcar use.

For Coleman, Alfred's proposal to buy the DuPont Company came at an auspicious time. Johnson Steel had been hit hard and shut down in the long depression of the mid-1890s. Arthur Moxham joined a steel company in Nova Scotia. Coleman moved to a modest house in Wilmington to try his hand at two separate but promising ventures in button-making and rifles, but these businesses also were languishing in the depression. Huddled with Alfred in his modest rented house, Coleman laid out the terms of his participation: first, he would be president of the new company; second, he would own half the stock; and third, cousin Pierre would have to join them. Alfred agreed. Coleman then picked up the phone in the hallway and called the man who had taken over a Johnson Steel subsidiary in Ohio — Pierre S. du Pont.

Pierre, so shy and quiet that the headmaster at his Philadelphia high school had nicknamed him "Graveyard," also had kept moving, though without the full-throttle bravado of his Kentucky cousin. Six years younger than Coleman and Alfred, Pierre had also gone to MIT but had stayed to receive his degree in chemistry in 1890.[1] Pierre returned to his father's company and a position at the powder mills during the new era of Eugene and Frank's tenure, but he soon grew disillusioned with their management. Throughout the 1890s he and Alfred commiserated over the backward condition of the labs at the Brandywine mills and at Carney's Point, where Pierre had joined Frank in 1892 in patenting a smokeless powder for shotguns. Despite this achievement, Pierre was discouraged by Frank's shortsighted planning. Freezing winter temperatures practically halted production at Carney's Point, where ice from guncotton drippings covered the floor. Pierre wrote his brother Belin, "It is all the same old trouble of not looking

POSTAL TELEGRAPH-CABLE COMPANY IN CONNECTION WITH THE COMMERCIAL CABLE COMPANY.

16.

JOHN W. MACKAY, President.
J. O. STEVENS, Sec'y. WM. H. BAKER, V. P. & G. M.

JOHN W. MACKAY, President.
ALBERT BECK, Sec'y. GEO. G. WARD, V. P. & G. M.

TELEGRAM

The Postal Telegraph-Cable Company transmits and delivers this message subject to the terms and conditions printed on the back of this blank.

Received at FEB 12 19

THE LORAIN STEEL COMPAN

(WHERE ANY REPLY SHOULD BE SENT.)

1B AU C 14 Paid.

Wilmington, Del. Feb'y 11th, 1902.

P.S.du Pont, Lorain, Ohio.

Expect to have meeting tomorrow night. Will wire you Thursday morning, I think yes.

T.C.du Pont.-7:50 A.

ahead and making provision for things that are bound to happen sooner or later."[2] Pierre looked to his future and spied some new opportunities taking shape out West.

When just a boy of seven Pierre had noticed that his mother's instructions at the local store, "Charge it on the book," eliminated the need for money, so he too used the magic phrase for several glorious days of "purchasing" his own treats. But when the accounts came due at the end of the month he learned "the difference between a ledger book and leger-de-main," as he later put it.[3] It was a sound economic lesson, but one that many industrialists at the time forgot. Pierre saw Johnson Steel borrow money to stay afloat, only to succumb to its creditors. Dependence on outside financing was a danger he was determined to avoid.

When Moxham and Johnson sold Johnson Steel to J. P. Morgan's Federal Steel Company in 1898, they asked Pierre to preside over the plant's liquidation and to pursue new investments for the company's one successful venture, a street railway company in Lorain, Ohio. Pierre hesitated, but decided to give the old DuPont firm one last chance. On January 26, 1899, he met with the partners, told them of his offer in Ohio and asked about his prospects with the company. If he had hoped that the possibility of his leaving would jar loose a counter-offer, he was disappointed. Only the junior partners, Alfred and Charles, encouraged him to stay. So as Coleman and Elsie came East, Pierre went West.

Now, three years later, Coleman once again was calling on his steady and reliable younger cousin, and Pierre once again answered the call, happy at the prospect of being back on the Brandywine. He and his assistant, an astute young bookkeeper named John Raskob, arrived in Wilmington as a blinding February snowstorm set in. Pierre and Coleman had two days together before the roads were clear enough for them to travel out to Swamp Hall to meet with Alfred. There was much to discuss before they met with their independent-minded cousin.

Coleman and Pierre had learned a great deal at Johnson Steel about running a large business, especially the importance of careful planning. It was not enough to make a quality product. One also had to look ahead, closely monitor the national economy and its effect on market conditions, and then make sure that production and inventory matched supply and demand. Accountants, clerks, secretaries and lawyers would be required as never before at the new DuPont, and researchers, too, to improve products and keep abreast of the latest scientific developments. Professional engineers would be needed to master the many new technologies crucial to sophisticated production. The company also would need salesmen and executives who understood and accepted the rules and constraints of working in a large organization. Change was desperately needed at DuPont, but neither Pierre nor Coleman knew how much change Alfred would accept.

The three cousins met at Swamp Hall on February 18 and discussed the tasks before them: making a rough assessment of the company's net worth, establishing financing and terms of the sale they would propose to the DuPont partners and specifying the duties of the new officers. Pierre wrote to his younger brothers that "the wheel of fortune has been revolving at pretty high speed on the Brandywine during the last week or two." He then added the droll assessment, "We have not the slightest idea of what we are buying, but in that we are probably not at a disadvantage as I think the old company has a very slim idea of the property they possess."[4]

It had already been agreed that Coleman would be president and concentrate on the company's overall direction. Alfred would put his mill experience to use as vice president in charge of black powder production. As treasurer, Pierre would apply the financial skills he had sharpened in the steel and street railway businesses. Although Pierre estimated DuPont's worth as closer to $15 million than the $12 million sale price named by the former owners, $12 million it would be, the cousins agreed. Instead of cash, however, $12 million worth of bonds were issued to the six former owners,

"I think Yes" telegraphed Coleman (left) to his cousin Pierre (middle) in response to the request to help take over the family business. One week later the meeting took place that decided the future of the company. Before the final decision, the cousins were in frequent telegraph contact.

OPPOSITE RIGHT: In the early 1900s DuPont explored the possibility of buying sodium nitrate fields in Chile so the company would own the raw materials needed for powder production. John J. Raskob (right), Pierre's close assistant for several decades, traveled by ship with Elias Ahuja to South America to investigate DuPont's options there.

DUPONT'S GUN POWDER
ON, DEL
built 1869.
ARTHUR LAMOTTE
DU PONT
MAGAZINE
AND AGRICULTURAL BLASTER
Vol. 1 JULY, 1913 No. 1
The Du Pont Army
Carrying the Powder to Perry
Exiling the Thawing Kettle
Du Pont Gelatin
Explosives as Life Insurers
Farming With Dynamite
The Portable Gun Club
Manufactured Leather in the Automobile Industry
E. I. du Pont d
ESTABLISHED 1802
DU PONT
DU PONT POWDERS
Have been known, used and recommended.
Since 1802
THEY MUST THEREFORE BE THE BEST.
1905
1802 DU PONT. 1900
WILMINGTON, DEL.
DU PONT
GENERATIONS HAVE USED DU PONT POWDER
DU PONT AMERICAN INDUSTRIES
How Many Hides Has A Cow?
Not Enough!
DU PONT FABRIKOID
The Ideal Upholstery Material—Superior to Coated Split Leather
DU PONT FABRIKOID COMPANY
WILMINGTON, DELAWARE
DU PONT
Du Pont
1901
Du Pont Smokeless
THE SPORTSMAN'S DELIGHT.
DU PONT MAGAZINE
AUGUST 1918
DU PONT MAGAZINE
Vol. 8 April 1918 No. 4
DU PONT MAGAZINE
VOL. 8 NO. 2
FEBRUARY 1918
THE NATION BUILDER
DU PONT
E. I. du Pont de Nemours Powder Co.
THE WORLDS LARGEST MANUFACTURER OF
EXPLOSIVES
GUN POWDER WORKS
TUNNEL AND MINING
SALTPETRE AND SODA
H. A. WELDY & Co.,
TAMAQUA, PA.

Throughout the 19th century, most advertising for DuPont explosives was done by independent sales agents who simply were making their presence known. In 1909 DuPont created an Advertising Division within the Sales Department to help influence the buying public. The company revived a faltering sporting powder market by popularizing trap shooting and created a new market by persuading farmers that the best way to clear their fields of rocks and stumps was to use DuPont dynamite. As the company diversified and began competing in an increasing number of markets, the distinctive oval trademark, devised in 1907, kept the company's identity before the public. *DuPont Magazine* began publication in 1913 to promote DuPont products.

OPPOSITE TOP: **As the company grew, so did its Wilmington headquarters, the tallest building on the left. Though Coleman wanted to move the firm to New York City, he compromised by agreeing to move instead from the secluded Brandywine Valley to downtown Wilmington.**

OPPOSITE MIDDLE and BOTTOM: **When the cousins assumed ownership of the company in 1902, the Brandywine River yards were a hodgepodge of aging facilities. Alfred pushed modernization, such as the new 1905 machine shop filled with modern gears and tools. It didn't get the chance to serve the company like its ancestors, for Hagley Yards ceased production in 1921.**

including Alfred, and $20 million in new shares were apportioned among the former and new owners. All parties were thereby guaranteed a stake in the new company's profits. DuPont passed to a new generation without a single greenback changing hands. On February 28 the lawyers filed the final papers that transferred ownership of the company. Only the lawyers required cash payment, so each cousin put up a third of the $2,100 in fees.

The transition was symbolically made at the Brandywine office the next day when Frank brought the morning mail in to Pierre, shook his hand, then walked out. Although he was surprised by the suddenness of Frank's farewell, Pierre was ready for his new responsibilities. He knew that DuPont was worth more than the purchase price, but he also knew that the value of a business could be further enhanced by efficient organization and management. Before DuPont's potential value could be realized, its new officers would have to consolidate and streamline the firm's many holdings scattered across the country into what Pierre envisioned as "the Big Company."

In the heyday of the Gunpowder Trade Association, DuPont and its main rival, Laflin & Rand, had reduced the threat of price wars not only by establishing trade rules but also by buying significant shares of each other's subsidiaries. What hurt one company, therefore, hurt the other as well. The result was a balanced share of the market for the two industry giants. However, the duplication of efforts created by these interlocking commitments stood in the way of the efficient business organization that Coleman and Pierre planned for the new DuPont. The finely calibrated, carefully coordinated gears of their management system would grind and stall unless they could completely control all their assets.

Of particular concern was the fact that DuPont did not own a single dynamite plant outright. Instead, it held shares in the Eastern Dynamite Company, a holding company set up by major East Coast producers in 1895 to do for the dynamite business what the GTA had done for powder. Laflin & Rand held three times as much stock in Eastern Dynamite as DuPont. To Pierre and Coleman, the solution was clear: DuPont must purchase the very firm that had recently come so close to owning it. Pierre realized that Laflin & Rand was in much the same state as DuPont had recently been in: its assets were undervalued, its owners were elderly and tired, and they might be persuaded to sell. Although Pierre urged Coleman to close a deal quickly, "lest our inspection might lead the owners to resurvey their property and retire from the commitment," Coleman spent months in tense negotiations before overcoming the objections of Laflin & Rand's owners.[5] Even Alfred had been kept in the dark during the prolonged, secret negotiations that closed in August 1902 with the purchase of Laflin & Rand for $4 million. The deal brought eight dynamite plants into the company and added 10 black powder plants to DuPont's 11. These, along with the two smokeless powder plants at Carney's Point and Pompton Lakes, New Jersey, made DuPont by far the largest explosives manufacturer in the nation.

The Laflin & Rand purchase came as a surprise to Alfred, who took it as an affront to his position in the company. Relations between Alfred and Coleman had begun to deteriorate only a few weeks after their purchase of DuPont. President Coleman hadn't seen the powder mills in years, and in March he decided to take a tour with Alfred. The tour opened a breach that widened over time. Coleman was no stranger to the muscled grime of heavy industry, but he was not prepared for the panoply of horse-drawn carts and water wheels that spread out before him like an aging industrial Gulliver, pinioned along both banks of the Brandywine by innumerable small ties to the past. Back at the office Coleman confronted Alfred, declaring brusquely that the mills should be closed. Moreover, he said, if DuPont were to be a serious player in the business world, it would have to move its offices to New York.

Alfred was appalled. He agreed that the mills had been managed inefficiently before, but now that he was free of Frank and Eugene's control, he promised to upgrade them. Pierre intervened and Alfred kept his authority over the mills. Pierre also backed keeping the DuPont offices in Wilmington but favored moving them downtown. Coleman accepted the compromise. Alfred kept his word, expanding the substitution of steam power for horse power wherever safety allowed, completely rebuilding the machine shop, enlarging the dam upstream, and renovating the workers' housing. He even repainted St. Joseph's Church and arranged for an extension of the Wilmington streetcar line out to the mill area. For his part, Coleman settled for the addition of two floors atop the Equitable Trust building at Ninth and Market Streets. Eight stories tall, this had been Wilmington's first "skyscraper" when it was completed in 1888.

The general contours of the company's future emerged clearly during these first few months. Like a rocket's launch pad, Alfred would one day be left behind. The energetic Coleman, like a rocket's booster stage, would carry DuPont to new heights, but he would fall away before the journey's end. Pierre, the careful planner and patient mediator, would guide the rocket into steady orbit.

The new owners were careful to assemble a competent crew, for without outside talent the Big Company would never get off the ground. There were several new positions to fill. The plan called for three production departments (High Explosives, Smokeless Powder and Black Powder) and six functional

departments (Sales, Military Sales, Legal, Purchasing, Development and Real Estate), each run by a vice president. Central administration operated through three committees: Administrative, composed of the eight vice presidents; Finance, headed by the company treasurer, Pierre; and the Executive Committee, which Coleman established in February 1903 to determine long-range goals and set broad policy.

Hamilton Barksdale, who had succeeded J. Amory Haskell as president of Repauno and had married Charles du Pont's sister, Ethel, was recruited to run High Explosives. Barksdale soon organized a research facility, the Eastern Laboratory, within his department. Francis I. du Pont, Frank's son, took charge of the Smokeless Powder Department, while Alfred continued as head of Black Powder. Arthur Moxham came from Nova Scotia to head DuPont's Development Department, which was responsible for monitoring competitors, tracking raw materials, and keeping abreast of relevant research. Moxham also organized DuPont's second research lab, the Experimental Station, near the old mills.[6] Haskell took over DuPont's Sales Department. These four men, along with the three du Pont cousins, formed the first Executive Committee during 1903.

Under Coleman and the Executive Committee, DuPont was a perpetual work in progress. Even after the purchase of Laflin & Rand, consolidation of the U.S. explosives industry remained a top priority. In the spring and summer of 1903, Coleman traveled to San Francisco to see whether he could purchase several additional competitors, including the large California Powder Works. The owners proved to be tough and cagey negotiators, however, and an exhausted Coleman checked into a hospital briefly to regain his strength and treat an abdominal pain. He wrote to Barksdale at the end of July, "If there is ever another job which includes as many different phases of as many different businesses and as many different characters as this one and anybody wants it, they need not consider me as a rival. With this crowd here unless a man is tricky and slick at the business he had just as well jump into the bay."[7]

Coleman did not jump. He returned to Delaware with the ownership of the California Powder Works. DuPont was now growing at a rapid clip. The company had acquired its largest competitors, along with their extensive holdings, while it continued to absorb numerous small powder firms. DuPont's Executive Committee decided to establish a holding company to help manage its many new acquisitions, and on May 19, 1903, the E. I. du Pont de Nemours Powder Company was incorporated in New Jersey for this purpose. Coleman and Pierre's Big Company carried on with further consolidation, picking up an astonishing 108 smaller companies by the end of 1907. Coleman paid a price for his whirlwind pace. He was beset by a variety of ailments, ranging from stomach pains to eyestrain that sometimes forced him to retreat into the sanctuary of a dark, quiet room. Coleman had surgery for removal of gallstones in 1909 and underwent abdominal surgery five years later, but doctors were unable to agree on a diagnosis of his various medical problems. Pierre grew more and more accustomed to standing in for him as DuPont's president.

The orders that came down from DuPont's new executives kept a growing corps of clerks, accountants and secretaries busy. In 1903 the company's office staff numbered 12 people. Just a year later it had grown to more than 200. When the cousins bought the company, DuPont did not have a single, full-time, salaried research scientist on its staff. A year later 16 research scientists were employed at the Eastern and Experimental Station laboratories. Science at DuPont ceased to be a matter of individual curiosity and invention and took on a new, institutional shape. The change was rapid. By 1911 the Experimental Station alone kept 36 chemists busy.

In 1902 Coleman appointed DuPont's Cincinnati sales agent, Robert S. Waddell, to direct sales operations. As with the rest of the new company's activities, teamwork and close organization replaced individual autonomy. Instead of

TOP LEFT: **Hamilton Barksdale, a du Pont in-law, was recruited to run the High Explosives division as part of the reorganization of 1902.**

TOP MIDDLE and RIGHT: **Rokeby Mill, built in the 1760s, was purchased by DuPont in 1850 for use as a machine shop, and then became the first Experimental Station.**

CENTER LEFT: **Pierre Gentieu at the Sportsmen's Exposition, Madison Square Garden, 1899.**

CENTER MIDDLE: **Experimental Station researchers pursued the development of explosives in DuPont's three divisions: black powder, smokeless powder and high explosives.**

CENTER RIGHT: **DuPont's growth demanded a larger headquarters office staff.**

BOTTOM LEFT: **Almost 100 years old in 1900, Hagley Yards was outmoded by newer, larger and more efficient plants elsewhere in the country.**

BOTTOM MIDDLE: **The new Hagley Yards machine shop in 1914.**

BOTTOM RIGHT: **Henry Clay Village on the Brandywine.**

G. Mathewson..........	66 years	W. Rowe, Jr.............	33 y
A. Burns................	63 "	J. A. McVey............	33
M. Foster..............	54 "	L. Ward................	33
			33
			33
			33
			32
			32
			32
			32
			32
			32
			32
			32
			32
			31
			31
			30
			30
			30
			30
			30
			30
D. Fisher..............	37		30
J. Stewart..............	37 "	J. McKenna.......	30
S. Frizzell..............	36 "	J. Farren................	29
A. Fleming..............	35 "	D. Dougherty...........	29
J. Maxwell..............	35 "	J. McLoughlin..........	29
B. Dougherty...........	34 "	P. McDade..............	29
D. Buckley..............	34 "	M. Campbell............	28

cutting deals on their own terms and offering rebates to their favorite customers, salesmen found that they now had to go by the company book. They received the latest technical training, filed standardized reports on set schedules, conformed to company pricing policies, and in general learned to march to a single drummer in Wilmington. Many prospered under such a system, but others grew resentful and resigned. Waddell himself found Coleman's tight ship too confining and left to set up his own powder company in Illinois. He was succeeded by J. Amory Haskell.

DuPont's many acquisitions helped increase the industrial workforce from about 800 in 1902 to more than 5,000 in 1910.[8] The huge scale and nationwide scope of its operations posed a challenge to the century-old legacy of personal management at the Brandywine mills. On the heels of the company's centennial celebration in July 1902, powderman Pierre Gentieu headed a delegation of old-timers who bid farewell to former leaders Frank, Henry Algernon, Charles and Alexis and then half-welcomed, half-warned the newcomers. "What the new company will do," Gentieu said, "we do not know, but let us hope that after one hundred years more, as much good can be said of them as is said today of the du Ponts of the last century."[9]

A number of powder workers, however, were unwilling to leave their fate to the good will and familial generosity of management. They turned instead to unionization as the key to their future security. DuPont had never been a union company. Indeed, the very assumptions underlying unionism offended the old-fashioned sense of shared interests, backed by firm control, that traditionally had characterized worker-manager relations at DuPont, notwithstanding the sporadic barn burnings of the 1890s. But in merging so many companies around the country after 1902, DuPont had acquired a few unionized plants, and organizers for the United Powder and High Explosives Workers of America soon came to the Brandywine.

The union chose poorly in targeting the home turf of

Resolutions

1802 1902

E. I. du Pont de Nemours & Co.

CENTENNIAL

JULY 4, 1902

At a meeting of du Pont Employees held at the Brandywine Club, July 9th 1902, the following Preamble and Resolutions were unanimously adopted.

Whereas, The Firm of E. I. du Pont de Nemours & Co. naturally proud of the record of One Hundred Years in the manufacture of Gunpowder in this country have celebrated the event in a fitting manner by holding a monster Picnic on their grounds on the Fourth of July 1902 and giving a public exhibition of Fire Works on the grounds of the Country Golf Club in the evening which thousands of persons came to witness thus adding general interest in the celebration of Independence Day.

Whereas, In thus celebrating their Centennial Anniversary the members of the

OPPOSITE: Coleman expanded the sales department — seen here in 1906 — and changed the way it operated, demanding that salesmen follow company pricing policies and file standardized reports.

TOP: The three cousins greatly expanded the support of scientific research in 1902. The Experimental Station employed 36 chemists by 1911. The entire staff of the station is pictured here.

LEFT: At the celebration of the company's 100th anniversary, a resolution was read on behalf of workers from the powder mills thanking the company for its kindness over the years.

Alfred I. du Pont. Alfred was a genial "father" to the workers and their families, dressing as Santa at Christmas time and sponsoring picnics and outings in the summer months. He would not brook the union's usurping his authority in the firm his great-grandfather had founded. When Local 130 struck the Brandywine mills in June 1906, Alfred set Pinkerton detectives to work, shut down the mills for two months, and fired 17 strike leaders, evicting them and their families from the workers' housing area and blacklisting them for any future employment with DuPont.

It was not through tough tactics alone that DuPont retained the upper hand in its struggle with the union. For example, Alfred co-opted one of the union's most talented and effective organizers, carpenter William Feenie, by promoting him to plant superintendent and then transferring him to

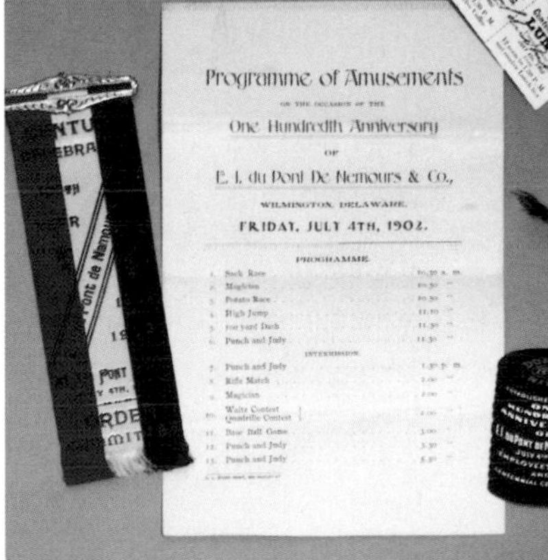

Programme of Amusements

One Hundredth Anniversary

E. I. du Pont De Nemours & Co.,

Wilmington, Delaware.

Friday, July 4th, 1902.

Programme

ABOVE: DuPont celebrated its 100th anniversary with festivities on July 4, 1902. The gala included fireworks, dancing and games.

NOTABLE

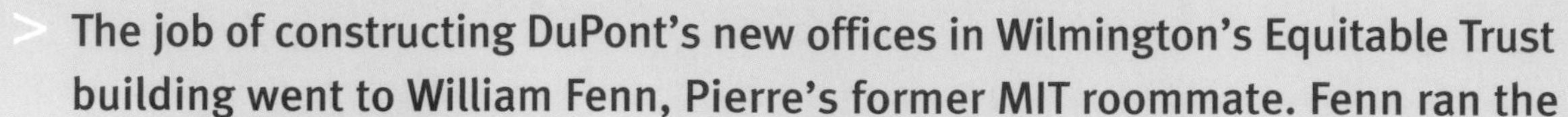

> The job of constructing DuPont's new offices in Wilmington's Equitable Trust building went to William Fenn, Pierre's former MIT roommate. Fenn ran the Manufacturers' Contracting Company, a DuPont subsidiary. The Eastern Dynamite Company had already occupied the fifth and sixth floors of the Equitable building, and with the purchase of Laflin & Rand in 1902, DuPont took over those floors. Fenn also added two new floors to the Equitable Building — a seventh and eighth — at Coleman's request. But DuPont soon outgrew these quarters, and in 1904 the Executive Committee authorized the construction of an entirely new building to house the headquarters staff of over 500 people. Typically, Alfred, Pierre and Coleman could not agree on naming the building, but Coleman realized people would end up calling it the "DuPont Building" no matter what they decided, so that is what the stone cutters carved over the front entrance.

> Why did DuPont establish two different research laboratories? This was a deliberate decision made by the Executive Committee, which settled on a two-pronged research and development strategy for the company. Researchers in the High Explosives Department's Eastern Lab were charged with working toward specific product improvements, such as developing non-freezing dynamite. Experimental Station researchers were given latitude to pursue projects of more general nature that could be of use to any of the three departments. Despite some conflicts and controversy, this led in time to a distinction between applied and basic research that proved highly successful at DuPont in later years.

> The early 1900s was an era of widespread "professionalization," with professionals of all kinds establishing credentials and forming associations. Chemists and engineers were among those who set ever more stringent standards for entry into their ranks, with degrees from a few respected American universities being considered appropriate credentials. But "professionalization" was an uneven process, and holdovers from the past remained — even at DuPont. Chemists like Arthur La Motte and Fin Sparre had been to college but had not graduated. Hudson Maxim (brother of the inventor of a prototype machine gun) received his chemistry training at seminary school. There were even self-taught independent inventors like J. N. Wingett, nicknamed "The Wizard" not only for his cleverness but because he liked to ponder problems in his "Spook Room," a small room painted completely black and furnished with only a black table and two black chairs.

The new DuPont company expanded not only by acquiring its competitors — 108 between 1902 and 1907 alone — but also by building new plants to meet specific market needs. In 1906 DuPont built its Louviers plant, south of Denver, to supply Colorado's burgeoning mines. Then the company began construction of DuPont, Washington, on Puget Sound to supply the Pacific Northwest. In 1912 the Hopewell plant was built on the James River in Virginia to allow ready access to Atlantic shipping. The Hopewell and Du Pont, Washington, sites grew dramatically during the First World War to become booming company towns.

Today, Pierre du Pont is widely considered to have been an organizational genius, but he drew his inspiration from unlikely sources. While considering various ways to realign DuPont's management in 1914, Pierre was taken by a chart he had seen that laid out the complex organization of New York's McAlpin Hotel, which Coleman had helped finance in 1910. At the center of the spoke-like diagram was the hotel manager. Superintendents of 16 departments, covering 78 functions, radiated outward from the manager, supplemented by 20 additional services such as barbers, florists, chiropodists, tailors, and the drug store and news stand. Comfortably ensconced within this beehive of activity was the Satisfied Guest.

In 1911 DuPont published *Tree Planting with DuPont Dynamite*, detailing "new and valuable information" for orchard growers. In this booklet, DuPont laid out the advantages of blasting holes for planting with dynamite rather than digging with a spade. According to the booklet, dynamite helped to loosen the earth all around, breaking up hard subsoils and allowing greater moisture access to the planted tree. One happy customer, the president of A. R. Bornot Bro. Co., prepared an orchard in just two days and reported that, "Three men could not have done the same amount of work in a week. The ground is now very loose; I am more than pleased and would not plant another tree on my place without explosives."

faraway Patterson, Oklahoma. But it was primarily through wage increases that Alfred won back the wavering allegiance of most of his wayward employees. As he wrote to one of his managers during the strike, there was nothing like "a judicious advance in wages" to remove workers' motivation to unionize.[10] Although Coleman had reservations about using wage increases to fight unions, returning workers were given raises and DuPont soon added a life insurance plan to the pension plan it had introduced in 1904 for employees with more than 15 years of service.

In 1906, Robert Waddell presented DuPont with a thornier problem than the union. He sued the company for treble damages, more than $1 million, blaming the failure of his Buckeye Powder Company on unfair trade practices by his former employer. Worse, Waddell carried armloads of damning internal documents from his DuPont days in Cincinnati to federal antitrust prosecutors. Rapid business consolidations such as those DuPont had engineered alarmed many Americans in the early 1900s. Whether it was in tobacco, oil, sugar or steel, they feared the power and influence of "the trust," which cartoonists inevitably portrayed as a menacing octopus with long, grasping tentacles. Ida Tarbell's 1903 account of Standard Oil in *McClure's Magazine* became a classic of investigative reporting, triggering a trend that Theodore Roosevelt disapprovingly dubbed "muckraking." Despite Roosevelt's effort to distinguish between "good trusts" and "bad trusts," the public demand for action could not be ignored.[11]

A legal means of dealing with these combinations had existed since 1890, although the Sherman Antitrust Act was seldom enforced during its first 20 years in existence. In one of the first of several big prosecutions, on July 31, 1907, the U. S. Justice Department filed an antitrust suit against DuPont. DuPont attorney James Townsend had seen it coming. He regarded the Gunpowder Trade Association not only as an impediment to consolidation but also as a legal risk. Townsend warned company officers that the GTA would likely be considered a "conspiracy in restraint of trade" and therefore in violation of the Sherman Act. In 1904 DuPont withdrew from the GTA, effectively dissolving it overnight, but the move was too little, too late.

Pierre had always been able to reconcile a sense of personal probity with the rough-and-tumble of the business world. For him, consolidation had been part of a progressive, forward-looking plan to increase efficiency and productivity, not an unfair undermining of competitors. Neither he nor any other DuPont officer, Alfred excepted, could fathom the Justice Department's assertions that the company had broken the law. Coleman was stunned for more mundane reasons: none of the strings he pulled in Washington could persuade Attorney General George W. Wickersham to call off the suit. Coleman professed illness to avoid testifying at the trial, so Pierre bore the brunt of defending the company on the witness stand. Unfortunately, Alfred's testimony only helped cement the government's case against DuPont.

This may have been due to Alfred's greater sensitivity to the political temper of the times, but it was certainly also a result of his changing position within the company. Alfred's attendance at Executive Committee meetings had always been spotty, and his influence waned so much that in 1911 Coleman, Pierre, Moxham and Barksdale restructured DuPont's management, placing all production under Barksdale, who was named general manager. Alfred was relieved from his responsibilities at the Brandywine mills and effectively exiled to a vice presidency on the company's Finance Committee. It was an important assignment but one that held little interest for Alfred, who vigorously protested to no avail. He remained isolated and alone in the upper ranks of DuPont's management, cut off from the mills and the workers who had so long sustained him. An emotional farewell ceremony in June 1911, in which the Brandywine workers presented him with a silver cup paid for out of their wages, seemed only to underscore his distance from his cousins and the new DuPont.

Alfred remained aloof from the legal arguments and the strategies hammered out in DuPont's executive ranks during the antitrust suit. Further, he showed no enthusiasm for the new company's acquisitions. But he was a member of top management, and the ignorance of DuPont's activities he confessed on the witness stand, combined with his generous volunteering of information about the company's complex operations, made Coleman and Pierre look devious and conspiratorial. Pierre was deeply embittered by the federal court's final ruling against DuPont on June 21, 1911, and by the remedy ultimately decided on after an additional year of acrimonious negotiations between the company and the Justice Department.

On June 12, 1912, DuPont agreed to divest itself of assets sufficient to create two new powder companies, Hercules and Atlas, and to turn over to them enough resources to ensure that they could produce 50 percent of the country's black powder and 42 percent of its dynamite. Further, DuPont agreed to share its research and engineering facilities with the two companies for five years.[12] But the court's medicine looked far worse than it tasted. Colonel Edmund G. Buckner, head of DuPont's Military Sales, successfully argued that quality control required a single supplier to the military, so DuPont retained all of its smokeless powder plants. The company also kept its huge Mooar mills in Iowa as well as its Brandywine mills, and managed to shift onto Hercules and Atlas several unproductive

LEFT: Colonel Edmund Buckner was head of military sales for DuPont during World War I.

BELOW: Ousted by his cousins in 1911, Alfred I. was sent off with a party thrown by the powder workers with whom he had worked closely for decades.

WESTERN UNION

WESTERN UNION

TELEGRAM

Form 260

GEORGE W. E. ATKINS, VICE-PRESIDENT NEWCOMB CARLTON, PRESIDENT BELVIDERE BROOKS, VICE-PRESIDENT

RECEIVER'S No.	TIME FILED	CHECK
		✓

SEND the following Telegram, subject to the terms on back hereof, which are hereby agreed to

Feb. 20, 1915

T. C. duPont,
Blackstone Hotel,
Chicago, Illinois.

Have arranged final conference with bankers Tuesday and Wednesday. Propose purchasing your sixty three thousand two hundred fourteen common at two hundred and thirteen thousand nine hundred eighty nine preferred at eighty, paying eight million cash and five million seven hundred sixty two thousand in seven year five percent notes of company to be formed. Collateral on notes to be thirty six thousand ~~hundred~~ shares common stock. We to have prvilege of paying notes before maturity. Also prepared to close immediately. Do you accept this proposition. Has Dunham full authority and necessary power to act for you. Important this be kept confidential for present.

Pierre

and outmoded plants. Ironically, the only clear loser when all parties broke free of the legal tangle that Robert Waddell had helped create was Waddell himself. He lost his personal suit in 1914 when a jury decided that he had created Buckeye Powder mainly to force a purchase offer by DuPont.

The prolonged antitrust battle should have alerted DuPont executives to the importance of public relations, but their righteous indignation blinded them to this valuable lesson. It took a series of deadly explosions in early January 1916 to impress upon Pierre that systematic and deliberate liaison work with the newspapers could work to the company's benefit by helping to dispel, rather than foment, rumors. After the explosions, R. R. M Carpenter, then director of DuPont's Development Department, recommended establishing a Publicity Bureau.[13] At the end of the month, the Executive Committee approved the idea, and DuPont took its first step toward shaping public opinion at the dawn of a new era of mass communications. The move came just in time, for the company and the family were soon mired in a public relations disaster. The myriad personal and business conflicts that had been brewing among Alfred, Coleman and Pierre roiled to the surface and were splashed over the front pages of America's newspapers for months. At the center of controversy, feeding the fracas with cold fury, was Swamp Hall orphan Alfred.

Coleman was a builder by temperament, not a patient manager. By 1913 he had grown weary of DuPont and looked to Manhattan for more of the sort of excitement he had experienced constructing the McAlpin Hotel, where he occupied a penthouse suite during his many New York visits. His heart was set on a new, 36-story skyscraper in New York City for the Equitable Life Assurance Society, which he now controlled, but by late 1914 the massive project had consumed his available capital and he badly needed more. In December, Coleman told Pierre he wished to sell his DuPont stock and asked him to sound out Alfred on the matter. Anyone who obtained Coleman's substantial holdings would have a large say in how DuPont was run, and Pierre's first instinct was to preserve control of the company. Pierre needed to contain that threat, but he also wanted to seize the opportunity it presented by getting control of his cousin's stock. He did not know how Alfred would react, but when Coleman entered the Mayo Clinic in mid-January for surgery, Pierre wound up mediating a fractious tug-of-war between his cousins that muddied not only the tuggers but the referee as well.

At first, Alfred's objection to DuPont's buying back Coleman's stock did not center specifically on who would own it. Instead, he declared that the $160 per share price Coleman and Pierre suggested in December 1914 was too high. Pierre and the rest of the Finance Committee replied that it was probably not high enough. After all, Allied war orders were raising the stock's value. By spring 1915 it was selling at $300 a share, though no one knew how long the fighting would last or when an armistice might leave the company with a large surplus of suddenly devalued powder, as had happened after the Spanish-American War. The risk and uncertainty of the war, as it then appeared, clouded rather than clarified the worth of Coleman's shares.

World War I had not yet bogged down in the trenches, but the negotiations between Coleman and the Finance Committee had. Pierre seized the opportunity that Alfred's obstinacy created. He and John Raskob quietly bypassed the Finance Committee and arranged a complex deal through J. P. Morgan that closed the sale for Coleman at $200 a share in February 1915. Most of the proceeds went to a new syndicate Pierre organized, DuPont Securities, later named Christiana Securities, instead of to the DuPont company itself. This was done in order to keep control of the older firm from being spread too widely among the stockholders.[14] Syndicate members were Pierre and his brothers Irénée and Lammot, their cousin Alexis Felix du Pont, John Raskob, who was now DuPont's treasurer, and R.R.M. Carpenter, head of the Development Department. Pierre extended an offer to family

When Coleman decided to leave DuPont, he and Pierre negotiated the sale of his stock. This telegram from Pierre to Coleman spelled out the terms of the sale. The secretive manner in which the sale was arranged enraged Alfred I. and caused a rift in the family.

TOP LEFT: **Old Hickory, a DuPont munitions plant in Tennessee, was one of the largest wartime plants in the country during World War I. DuPont built 1,112 buildings for the factory and 3,867 structures for the adjacent employees' community.**

TOP RIGHT: **The Old Hickory site granulated its first powder 116 days after groundbreaking, 121 days ahead of the contract schedule.**

BOTTOM LEFT: **In November 1918, with World War I at an end, 12,500 men moved through the pay booths at Old Hickory in five hours to receive a final paycheck and a layoff notice.**

BOTTOM RIGHT: **DuPont built its Louviers explosives plant south of Denver, Colorado, in 1906 to supply the mining industry.**

CENTER: **Following the war, with Coleman and Alfred having left DuPont, Pierre ascended to lead the company. Coleman wrote to Pierre April 21, 1919, congratulating him on his election to chairman.**

shareholders to exchange their DuPont stock for the syndicate's stock. Those closest to the company's daily management generally accepted, but those least familiar with its operations tended to sympathize with Alfred and reject Pierre's offer. In Alfred's eyes the DuPont Company, not the new syndicate, best represented the family's true interests. But in a few days Pierre became the new president of DuPont, and six months later, in September 1915, the "DuPont Powder Company" of New Jersey re-incorporated in Wilmington under its old name, E. I. du Pont de Nemours and Company.

Again, Alfred was surprised and angered by a decision Pierre and Coleman had made without him. He had allies on the margins of the DuPont Company, however, and on December 9, 1915, his cousin Philip filed suit in federal court to block the sale of Coleman's stock to DuPont Securities. In April 1917, just days after the United States entered World War I, Federal Judge J. Whitaker Thompson ruled for Alfred's side. Pierre had misrepresented the actual situation, said Judge Whitaker, in telling Coleman that Alfred was rigidly fixed on a $125-per-share purchase price. Whitaker believed Alfred's contention that he had been willing to negotiate further with Coleman but had been preempted by the DuPont Securities arrangement, which Pierre had concluded without proper approval by the Finance Committee, of which Alfred was a member. But Judge Thompson decided to put the final resolution of the matter to a stockholder vote, and on October 10, 1917, after four years of high wartime earnings, DuPont's very satisfied shareholders voted their overwhelming support for Pierre, DuPont Securities and the status quo.

The lawsuit was a moral victory of sorts for Alfred, but it effectively ratified Pierre's actions. Moreover, it cost Alfred his vice presidency and his seat on DuPont's Board of Directors. Coleman's determination to answer the siren call of Manhattan cost him dearly, too. The $14 million he was paid for his stock would have been $58 million if he had waited just six months longer. By October 1915, DuPont stock had risen to $900 a share, four-and-a-half times what DuPont Securities had paid Coleman. In 1913 DuPont's net profits had totaled nearly $5.5 million. In 1915 they were almost $57.5 million, more than a ninefold increase. Profits for 1916 rose still higher to $82 million, surpassing the combined profits of every year since 1902, when Coleman, Pierre and Alfred had bought the firm.[15]

These profits resulted not only from the unexpected duration and intensity of the war but also from farsighted business practices. DuPont's contracts to supply the European Allies in the years before America's late entry into the conflict were written so as to recover all of the company's costs for plant construction and expansion. Originally intended to protect shareholders from great loss if the war ended and the Allied contracts were canceled suddenly, the policy succeeded so well that DuPont could actually lower its prices in 1916. DuPont's yearly production capacity had approached 8.5 million pounds of explosives in 1914. At the war's end four years later, it was more than 10 times larger — half a billion pounds, much of it coming from huge, newly constructed plants at Hopewell, Virginia, and Old Hickory, near Nashville, Tennessee. The 1.5 trillion pounds of explosives supplied by DuPont — 40 percent of the total used by the Allies — saved the British Army, said the chief of the British Munitions Board.[16]

Though DuPont was prospering mightily, its public image was faring less well. Americans were deeply divided over the war before weighing in for the Allies in 1917. Some people believed fervently in the Allied cause, while others wished to steer clear of the bloody implosion of Europe's corrupt old regimes. Despite Pierre's efforts to keep DuPont above the political fray, the company was ineluctably drawn in as peace advocates and German sympathizers accused DuPont of camouflaging its greed for profit by supporting "preparedness" campaigns. Even after America's entry into the war, waves of patriotism and anti-German propaganda did

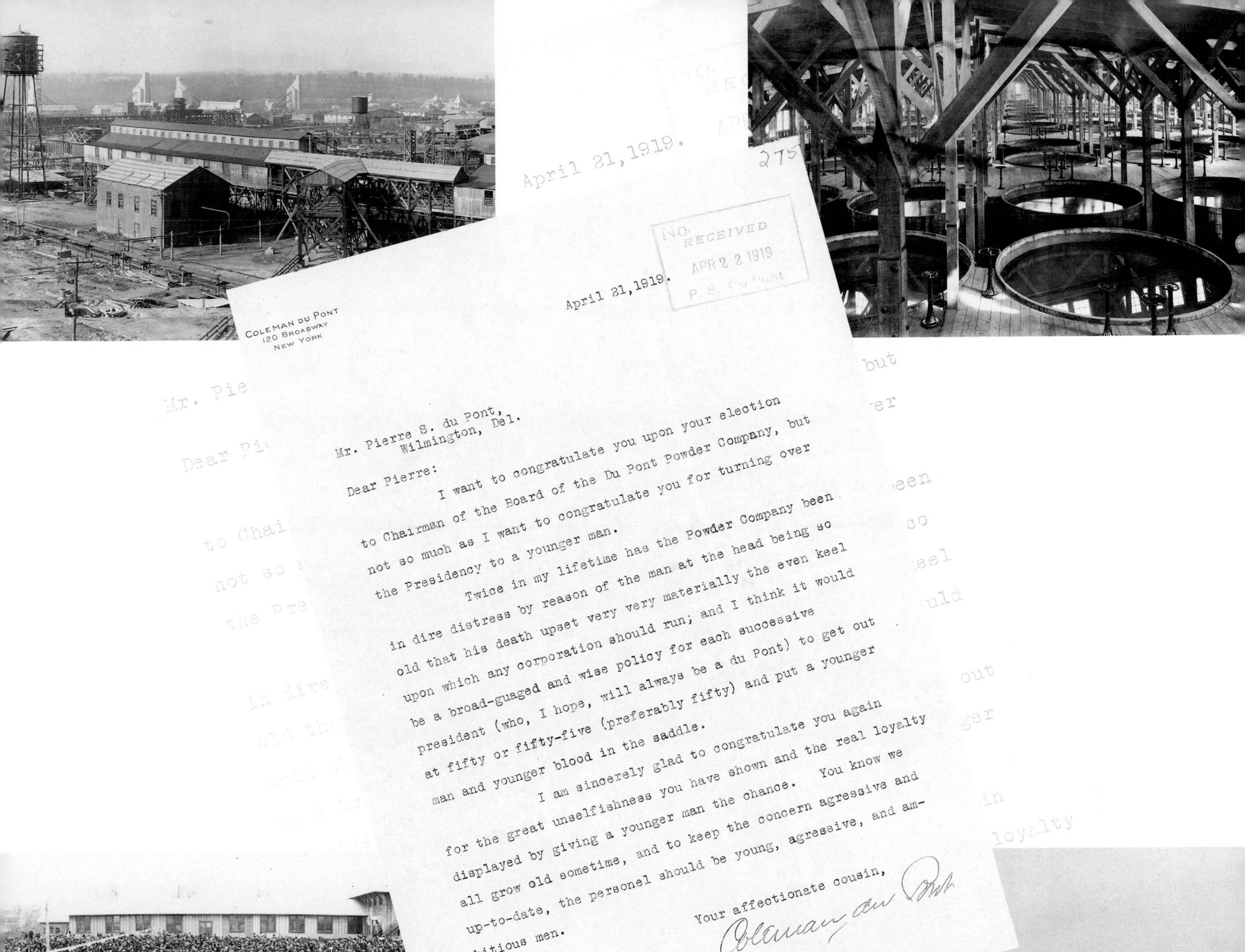

COLEMAN DU PONT
120 BROADWAY
NEW YORK

April 21, 1919.

No. RECEIVED
APR 22 1919
P. S. du Pont

275

Mr. Pierre S. du Pont,
Wilmington, Del.

Dear Pierre:

I want to congratulate you upon your election to Chairman of the Board of the Du Pont Powder Company, but not so much as I want to congratulate you for turning over the Presidency to a younger man.

Twice in my lifetime has the Powder Company been in dire distress by reason of the man at the head being so old that his death upset very very materially the even keel upon which any corporation should run; and I think it would be a broad-guaged and wise policy for each successive president (who, I hope, will always be a du Pont) to get out at fifty or fifty-five (preferably fifty) and put a younger man and younger blood in the saddle.

I am sincerely glad to congratulate you again for the great unselfishness you have shown and the real loyalty displayed by giving a younger man the chance. You know we all grow old sometime, and to keep the concern agressive and up-to-date, the personel should be young, agressive, and ambitious men.

Your affectionate cousin,

Coleman du Pont

E.

HOTEL
THE WILMINGTON BARBER SHOP.
VIRGINIA
LUNCH
CENTRAL HOTEL
HOPEWELL St.
AUG 8th 1915
HOPEWELL VA.

A KEY INGREDIENT OF SMOKELESS POWDER WAS NITROCELLULOSE. AT THE OUTBREAK OF WAR IN 1914, DUPONT HAD FACILITIES FOR MAKING ABOUT 1 MILLION POUNDS OF IT A MONTH. BY WAR'S END, OUTPUT WAS MORE THAN 1.5 MILLION POUNDS A DAY, MOST OF IT FROM A SPRAWLING NEW COMPLEX IN HOPEWELL, VIRGINIA. FACILITIES WERE BUILT FOR MAKING SULFURIC AND NITRIC ACIDS, FOR PURIFYING COTTON LINTERS, AND FOR CONVERTING THESE INGREDIENTS INTO NITROCELLULOSE. HOPEWELL VILLAGE WAS BUILT TO HOUSE THE 28,000 EMPLOYEES AT THE PLANT, MOST OF WHOM CAME

TOP: **The Old Hickory facility near Nashville, Tennessee, was the largest construction project of World War I. Powder was produced three months and nine days after the final contract was signed to begin construction.**

MIDDLE: **A cartoonist portrayed DuPont's impact on the Kaiser.**

BOTTOM: **With the nation's men fighting in Europe, women staffed factories. Women take their lunch break here at DuPont's Newburgh, New York, September 14, 1918.**

OPPOSITE LEFT: **As war raged in Europe, DuPont was swamped with Allied munition orders. On October 12, 1914, alone, France ordered 8 million pounds of cannon powder and 1.2 million pounds of guncotton.**

OPPOSITE TOP: **At Brandywine Mills, "Bloomer Girls" load Balistite rings — shell propellants — in June 1918.**

OPPOSITE MIDDLE: **Powder that served as a propelling charge for the 3-inch Stokes Trench Mortar was packed in silk bags.**

OPPOSITE BOTTOM: **Caps and fuses were made at Pompton Lakes, New Jersey. Peak daily output reached 1.5 million caps.**

not wash away lingering suspicions about the power of the powder trust and about the machinations of what Theodore Roosevelt had once called "predatory wealth." Congress expanded the three-year-old federal income tax to finance the war effort, placing most of the load on wealthy individuals and corporations. The "excess profits" tax, as it was known, fell most heavily on DuPont and other producers of war-related goods, which paid more than half of all federal taxes collected in 1918.

Uproar surrounding DuPont's mammoth Old Hickory plant reflected a long-standing ambivalence within the American public about the virtues and vices of big business. The plant's namesake was Andrew Jackson, the pugnacious Tennessean who rode a wave of populist resentment against Eastern money to the White House in 1828. Nearly 90 years later, government officials like Secretary of War Newton Baker attempted to make political hay by similarly condemning DuPont's power and influence. During contract negotiations in the fall of 1917, Baker charged DuPont with war profiteering and claimed that the federal government could construct its own explosives plant. The American people could "do things for themselves," he insisted.[17] After some failed attempts in that direction, however, Baker reconciled himself to dealing with "the trust." He signed the contract with DuPont for Old Hickory on January 29, 1918. Ten days later ground was broken, and 4,700 acres of farmland rapidly disappeared under what became the largest smokeless powder plant in the world.

DuPont's Engineering Department organized the construction of an industrial city of 35,000 people, which included schools, churches, stores, banks, fire and police departments, railroad facilities, paved roads, a 400-foot trestle bridge and a YMCA. Old Hickory's water system could easily have supplied Boston or Cleveland. Its electrical generating capacity exceeded all of nearby Nashville's needs, and its refrigeration plant was the world's largest. At their peak in

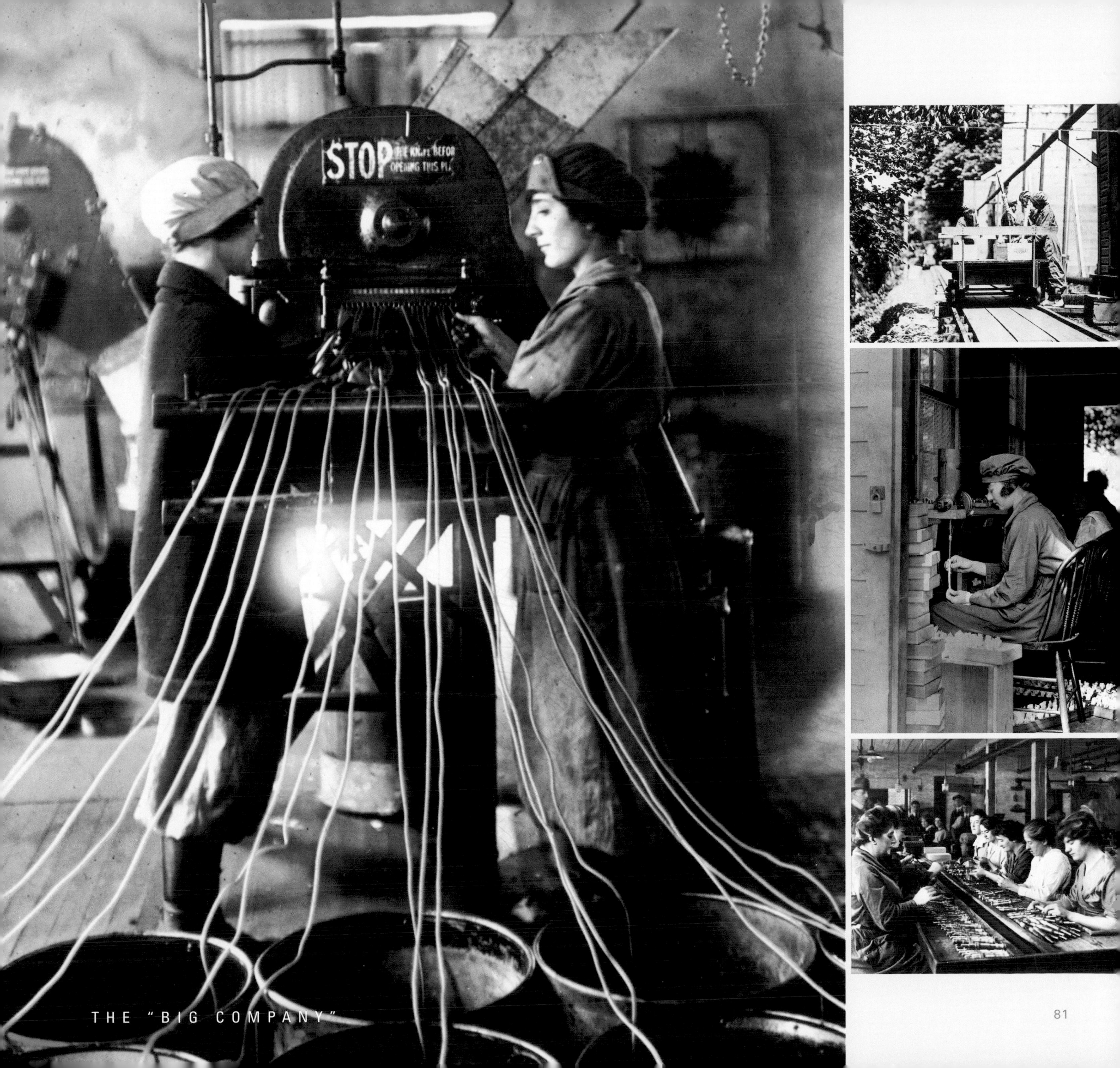
STOP THE KNIFE BEFOR
OPENING THIS PL

DuPont bought the Fabrikoid Company of Newburgh, New York, in 1910, signaling the purchase of established firms — complete with plants, patents and trained personnel in place — as the route to future DuPont expansion into new businesses. Fabrikoid was a waterproof fabric made by treating cotton cloth with successive coatings of a pliable nitrocellulose lacquer. It had been sold since 1895 as an artificial leather. DuPont chemists improved the product, making it tough, pliable and resistant to grease, oil, perspiration, mildew and other agents that cause leather to deteriorate.

Nitrocellulose plastics — called Pyralin — came to DuPont with the purchase of the Arlington Company.

August, Old Hickory's mess halls served 1,125,945 meals.

Old Hickory produced its first powder on July 2, 1918, in less than half the time called for in the contract, but the sixth of nine production units had just come on line when the Axis surrendered in November. Between 1915 and the November armistice, DuPont's employee ranks had swollen from 5,500 to 185,000. In the ensuing weeks and months of demobilization, DuPont laid off 90 percent of its company-wide workforce and Old Hickory became a ghost town. DuPont's net profit from the plant was less than half a million dollars, well under one-half of 1 percent of its $130 million construction outlay. With Old Hickory operating under the scrutiny of more than a hundred government auditors, Pierre was determined to leave no opportunity for criticism.[18]

Pierre knew the risks as well as the opportunities of war production. He also absorbed quickly the many public relations lessons the Big Company had taught him. Explosives had become a risky business in more ways than one. The antitrust suit, along with the war profits charges, suggested that diversification might be a prudent strategy for DuPont. From its beginnings under Coleman in 1903, the Executive Committee had explored alternative uses for the company's nitrocellulose products. In 1904 DuPont had purchased the International Smokeless Powder and Chemical Company in Parlin, New Jersey, thereby acquiring that firm's brass lacquer and solvents businesses. In 1910 it had bought the Fabrikoid Company of Newburgh, New York, a producer of artificial leather. DuPont bought the Arlington Company, a maker of nitrocellulose plastics, the Fairfield Rubber Company and the Harrison paint company in quick succession between 1915 and 1917. These acquisitions had provided DuPont with a great deal of practical knowledge. The Great War had given the company a great deal of money — nearly $310 million. The company soon tapped into these deep reservoirs of capital and scientific expertise for a new purpose — serving American consumers.

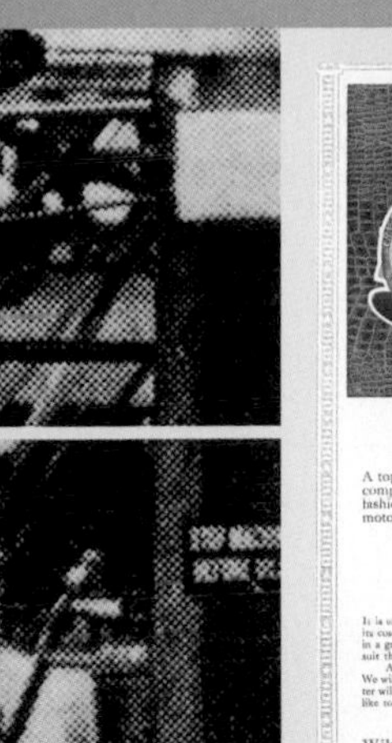

DU PONT AMERICAN INDUSTRIES
Tailor-Made Tops
of Craftsman Fabrikoid
A top of Craftsman Fabrikoid has a striking elegance and individuality that compels admiration. It gives, more surely than anything else, that ultra-fashionable character—that super-quality appeal so sought after in fine motor cars.
Craftsman DU PONT FABRIKOID Quality
Is the Aristocrat of Top Materials
It is entirely different and distinct from all other top materials, and, although the highest priced, its cost is more than justified by its de luxe quality, exceptional beauty and long life. It is made in a great variety of grains and colors (solid and Moorish) to harmonize with any finish and to suit the taste of any car owner.
Any expert top maker can put on a victoria or semi-victoria top of Craftsman Fabrikoid. We will send small samples of the material on request. Their appearance and distinctive character will delight you. Write for samples and prices today, telling us some of the colors you would like to see.
Du Pont Fabrikoid Company, Inc.
Wilmington, Del.
New Toronto, Canada
DU PONT

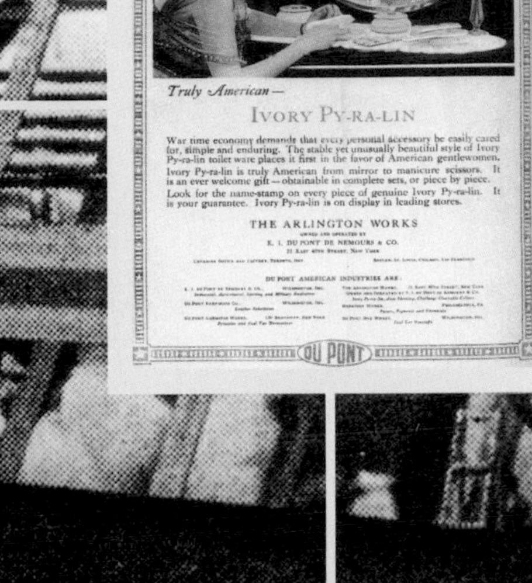

DU PONT AMERICAN INDUSTRIES
Truly American—
IVORY PY-RA-LIN
War time economy demands that every personal accessory be easily cared for, simple and enduring. The stable yet unusually beautiful style of Ivory Py-ra-lin toilet ware places it first in the favor of American gentlewomen.
Ivory Py-ra-lin is truly American from mirror to manicure scissors. It is an ever welcome gift—obtainable in complete sets, or piece by piece.
Look for the name-stamp on every piece of genuine Ivory Py-ra-lin. It is your guarantee. Ivory Py-ra-lin is on display in leading stores.
THE ARLINGTON WORKS
E. I. DU PONT DE NEMOURS & CO.
DU PONT AMERICAN INDUSTRIES ARE:
DU PONT

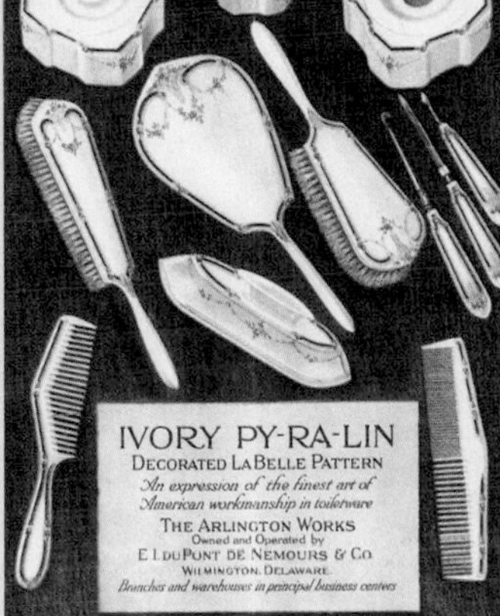

IVORY PY-RA-LIN
DECORATED LA BELLE PATTERN
An expression of the finest art of American workmanship in toiletware
THE ARLINGTON WORKS
Owned and Operated by
E. I. DU PONT DE NEMOURS & CO.
WILMINGTON, DELAWARE
Branches and warehouses in principal business centres

CHAPTER 4 SERVING THE NEW CONSUMER

RIGHT: **William C. Durant, founder of General Motors.**

FAR RIGHT: **As DuPont emerged from World War I, it was still a family company. Trusted aides such as John Raskob were admitted to the fold, but the du Ponts filled most leadership positions through the 1920s. This was the 1919 Finance Committee.**

LETTER: **"I recommend that our company purchase Chevrolet," wrote John Raskob, DuPont treasurer, to the Finance Committee September 19, 1917.**

BOTTOM: **DuPont became a major corporation in the 20th century and established committees and vice presidents to help run the growing organization. The Executive Committee made all major decisions.**

Copy

December 19th, 1917.

TO FINANCE COMMITTEE
FROM TREASURER

GENERAL MOTORS-CHEVROLET MOTOR STOCK INVESTMENT

I recommend that our Company purchase Chevrolet Motor Company and General Motors Company common stocks in accordance with plan herein outlined and with a view to bringing this formally before our company I have asked for a special meeting of the Finance Committee to be followed as soon as possible by a special meeting of the Board of Directors to consider the following resolution:

RESOLVED that the President and Treasurer of this Company be and they are hereby authorized to purchase up to $25,000,000.00 worth of the common stocks of the Chevrolet and General Motor Companies paying for the Chevrolet Motor Company common stock an average price not to exceed $115.00 per share and for the General Motors Company common stock an average price not to exceed $95.00 per share; and

BE IT FURTHER RESOLVED that they be and are hereby authorized to do all acts and things necessary to finance and carry out this purchase in accordance with plan outlined in Treasurer's report to Finance Committee, dated December 19th, 1917.

"It is easier to make war than to make peace," said France's premier Georges Clemenceau in 1919 after the great guns of the World War had been silenced. The headaches of Versailles were better than the heartaches of Verdun, but peace, like war, brought its own share of difficulties. DuPont's leaders also had a clear preference for the challenges of peacetime over the demands of war, despite the profits that war production generated. But those challenges were not easy, nor were they easily met.

The company's assets had quadrupled during four years of war. Gross receipts in 1912, just before the court-ordered dissolution of DuPont's black powder and dynamite businesses, totaled $35 million, about $600 million in today's dollars. It had been the company's best year before the war. In 1918, gross receipts reached $329 million. After paying out dividends in 1918, DuPont posted accumulated earnings of $68 million, more than seven times its 1915 amount.

Assets could not be measured only in dollars, however. The number of trained executives at DuPont nearly tripled during the war, from 94 to 259. The Engineering Department grew from 800 to 4,500 men. Fresh from numerous wartime building and expansion projects, this group of highly skilled experts, along with the facilities they had constructed, represented large investments that DuPont was loath to lose. DuPont had to find peacetime outlets for its expanded physical and personnel resources. The general answer was clear enough: translate the chemistry of explosives into the chemistry of the consumer market. But the specific ways to do this were anything but obvious. What products should be developed? What new expertise would be needed? Would the company have to modify its basic structure? How much would it all cost?

DuPont responded by taking a series of carefully calculated risks. The Finance Committee accepted treasurer John Raskob's advice and in January 1918 approved a $25 million investment in General Motors, a company whose future success seemed likely but by no means certain under the mercurial leadership of founder William C. Durant. A heavy investment in GM, Raskob informed the Finance Committee a month earlier, "will undoubtedly secure for us the entire Fabrikoid, Pyralin, paint and varnish business of those companies [General Motors-Chevrolet]."[1] Both Raskob and president Pierre S. du Pont had bought into GM as early as February 1914, and in 1915 Pierre had been invited to join GM's Board of Directors.

It was hardly unusual for DuPont's Finance and Executive Committees to decide that what was good for the company president and treasurer would be good for the company. But in January 1918 DuPont was in the middle of contentious negotiations with the federal government over building the Old Hickory plant in Tennessee. Investing in GM meant committing a quarter of DuPont's available investment capital to an unfamiliar industry just when that money might be needed to meet what would be the largest wartime contract ever awarded in America up to that time. Despite concern expressed by some members, DuPont's Board of Directors accepted the unanimous recommendations of the Finance and Executive Committees, approving the GM stock purchase on December 21.

TOP LEFT: **Fin Sparre served in numerous posts for the company, including Director of the Development Department. He also became a member of the Board of Directors.**

TOP RIGHT: **Charles Reese established the first DuPont research labs at Repauno in 1902, and later became Director of the Chemical Division as the company expanded its research facilities.**

BOTTOM: **In its first years, the Experimental Station consisted of several unconnected buildings scattered around the grounds. By 1920, the complex had expanded as DuPont made a commitment to scientific research as a basis for its products.**

Pierre and the Executive Committee also accelerated DuPont's prewar efforts to spread beyond explosives into fields like dyestuffs, plastics and paints. The company was prepared to endure some short-term losses to realize future profits from the long-term development of new products. But a severe postwar recession increased the risks for DuPont. When the Wilson administration cut federal spending and the Federal Reserve tightened the money supply in a combined effort to slow inflation in the 1919 postwar spending boom, the economy went into a tailspin. The recession forced layoffs of valuable personnel, delayed consumer demand for the company's fledgling new products and raised doubts about DuPont's recent research and development expenditures. Even GM's prospects were sinking rapidly beneath mounting debts and internal management crises, threatening to render DuPont's sizeable investment worthless. DuPont's explosives departments posted a $2.5 million profit in the first half of 1921, but its diversified products lost nearly $4 million.

Nor could the DuPont executives and advisors who had worried over the GM investment late in 1917 take comfort from the company's experience with dyestuffs. From the Development Department's tentative explorations in 1915 to the mid-1930s, when DuPont at last recouped its total dyestuffs investment, dyes proved to be one of the most difficult — yet also crucial — ventures the company ever undertook. The Allied blockade of German exports in World War I had created a huge demand for domestically produced dyes as well as an opportunity for DuPont to enter a new business. America, the world's second largest consumer of dyes after China, quickly became "the most dye-starved nation in the world."[2] The chemistry of synthetic dyes and the "intermediates" used in making them was closely related to the chemistry of high explosives. Further, since these dyes were derived from coal tar as well as from nitrated cellulose, their development held tantalizing possibilities for new discoveries in the general field of organic chemistry. The potential of such research for improving existing products and bringing new ones to market was not lost on DuPont's scientists.

As early as the summer of 1915, Development Department director Fin Sparre and Chemical Department director Charles Reese argued strongly in favor of entering the dyestuffs field. The risks were great: DuPont lacked the expertise that had earned German firms a monopoly in the field for several decades. But the risks of not entering the market were equally high — an angry outcry from American consumers if DuPont should be found selling chemical intermediates to British dye producers or, worse, to German firms that might use them for war purposes. DuPont's Executive Committee contemplated the grim possibility of charges that it had put war profits ahead of the needs of the American consumer.

Historians David Hounshell and John Kenly Smith aptly describe the fitful manner in which DuPont entered the dyestuffs field: "Much like the unsupervised child that wades progressively further from seashore until suddenly the bottom drops out, DuPont waded deeper and deeper into dyestuffs."[3] Early in 1916, DuPont's own Fabrikoid and Pyralin production plants reported shortages of dyes. In response, DuPont's Experimental Station and Eastern Laboratory both took up dye research. On November 30 the company also contracted to purchase formulas and methods from a British producer, Levinstein, Ltd., that was producing indigo dyes based on confiscated German patents. A team of DuPont chemists and engineers spent two months in England studying Levinstein's operations and making detailed notes. Their work was cut short when Germany threatened to torpedo any ship found in British waters after February 3, 1917. With life jackets at the ready, the DuPont team steamed for home on February 2.

Lammot du Pont, whose Miscellaneous Manufacturing Department oversaw the dye venture, reported to the Executive Committee in August that expenditures to date on dyes amounted to nearly $9 million. Those expenses increased

when DuPont engineers built the Jackson Laboratory at Deepwater, New Jersey, in the spring of 1917. Within two years the company had spent more than $1 million at the new lab. Despite their new facilities, researchers soon became frustrated by unanticipated problems in producing particular hues consistently, then in scaling up production to turn out bulk quantities. Chemical engineers, who knew that there was more to replicating small-scale, laboratory successes than simply making all the equipment bigger, designed midsize production facilities called "semi-works" to test the viability of mass producing the new dyes.

The Armistice in November 1918 further complicated DuPont's prospects by raising the possibility of Germany's re-entry into the American dye business. German firms had taken out dye patents in the United States prior to the war, largely to block American competitors. But when America entered the war, Congress passed a Trading with the Enemy Act that allowed the government to seize, among other things, German patents and place them under the control of the Alien Property Custodian, A. Mitchell Palmer. Now that the war was over, returning the patents to German firms could undermine the dyestuffs operations of DuPont and other domestic dye producers. In January 1919 DuPont General Counsel John Laffey and other representatives of domestic dye producers met with Palmer's staff to discuss the disposition of the seized German patents. They decided to form and fund a nonprofit organization, the Chemical Foundation, Inc., which bought 5,700 patents from the Alien Property Custodian for $271,000.[4] The patents provided little help to the struggling American dye producers. They did nothing to solve the problems of large-scale production, and many of the patents actually contained false information that had been deliberately inserted years before by the German firms to throw future competitors off track.

A few months later, in the spring of 1919, a DuPont representative in Europe reported that a German chemical

OPPOSITE TOP LEFT: In the analytical section of the Deepwater, New Jersey, technical lab, DuPont tested dyes for several decades to improve their color quality.

OPPOSITE TOP RIGHT: Checking dye shades at Chambers Works, Deepwater.

OPPOSITE BOTTOM: In the dyestuffs lab, Dorothy Bennett mixes lake colors on a lithographer's stone to ensure conformity.

LEFT: The Deepwater dyeworks — renamed Chambers Works in 1944 to honor Dr. Arthur D. Chambers — was the birthplace of the American dye industry. The new dye business marked DuPont's entrance into the field of organic chemistry.

NOTABLE

> It's a long way from high explosives to hosiery. Though the products may seem similar to chemists, the general public needed a sustained marketing campaign before they would associate DuPont's brand name with consumer goods. Realizing that homemakers would be the main purchasers of many of its products (such as Duco paint, rayon hosiery and Pyralin toiletries), DuPont began to target its advertising directly to middle-class women. In the 1920s and 1930s, the company started to hire interior decorators, writers for women's magazines such as *Good Housekeeping*, home economists and other domestic advisors as spokespersons for its products. In 1929, decorator Agnes Foster Wright appeared in a DuPont advertisement claiming that the kitchen "offers more interesting possibilities to the woman who knows what can be done with Duco and a little ingenuity to create charming interiors." Well-known home economist Christine Frederick also recommended an exuberant use of Duco paint in the bathroom. "The mirror frame and medicine cabinet, the glass-holder, the towel-racks, why not colored instead of white? As soon as each piece was treated with my brush, a feeling of individuality, brightness and glow pervaded the bathroom. Then I set to work on the faucets, even the soap-dish and wire sponge basket. Laugh if you like. But the bathroom, which had been previously dull and even clammy, became gay and cheerful. Yet it took only a pint of lacquer to make the change." In the 1930s DuPont continued its corporate sponsorship of domesticity when the Rayon Department hired Emily Post, author of *The Personality of a House* as well as an etiquette expert, to host its Cellophane radio program and deliver home decorating advice to 13 million listeners.

> DuPont's Jackson Laboratory at the Chambers Works was named for chemist Oscar R. Jackson (1855–1916), who became superintendent of the Repauno dynamite plant in 1884. In 1890 Jackson and his assistant, James Lawrence, perfected the acid recovery process in nitroglycerine production, an effort that had claimed the lives of Lammot du Pont and four other men six years earlier. Jackson's technical advice was widely sought and valued, for he never ventured an opinion if he had any doubts about it. Jackson's father, Dr. Charles T. Jackson, discovered the anesthetic use of ether in the 1840s and was the first physician to use it in surgery.

> DuPont directed its attention southward after the turn of the century. In 1904, the company supplied explosives for the construction of the Panama Canal. At the request of the Brazilian government, DuPont built an explosives factory in that country in 1908. And in 1910, the company bought the near-bankrupt Oficina Carolina nitrate mine in Chile, renaming it "Oficina Delaware."

> DuPont established the Medical and Welfare Department in 1915 to promote good health within its workforce. The company required routine medical examinations for all employees, with additional, special examinations for those who regularly handled toxic chemicals.

> Americans found new ways to have fun in the 1920s. DuPont jumped at the chance to sell people the accessories they would need to participate in all of the new leisure activities, whether they wanted to listen to phonograph music or radio programs, play golf or tennis, or go to the movies, the beach or a college football game. Trying to find markets for two of its major products of the 1920s, Fabrikoid (synthetic leather) and Pyralin (plastic), DuPont produced a dizzying number of different kinds of items meant to help Americans enjoy themselves. For the beach, the company manufactured rubberized beach balls, beach shoes and rafts. To encourage the development and success of college athletics, DuPont manufactured enormous rubber covers for stadiums to protect football fields from rain and special Fabrikoid seats for the stands. The company made tea doilies of Fabrikoid as well as cases for phonographs, golf clubs, ice cream containers and overnight bags, and Pyralin radio dials and instrument mutes. The product "Nemoursa," fancier than Fabrikoid, found its way into bridge scorecard sets and pocketbooks. DuPont searched for markets that would best utilize its inventions but was willing to make its products available to every possible constituency. DuPont marketers suggested that archaeologists use DuPont cement to glue together ancient bones and artifacts and recommended the use of DuPont rug anchors to prevent household accidents. DuPont made Fabrikoid seat covers for the League of Nations and Pyralin leg bands for chickens. By experimenting with every conceivable application for its products, DuPont eventually found a niche in the consumer revolution.

> Chemical engineers transform small-scale, laboratory processes into full-scale manufacturing operations. Their expertise has always been central to DuPont's success. Chemical engineers established their own professional society, the American Institute of Professional Engineers, in 1908, but it took seven more years for one of them, Arthur D. Little, to formally describe the profession's theoretical underpinnings. Little's insight was that all chemical processes involve the same basic "unit operations": heating, cooling, distilling, crystallizing and drying. In the 1920s, Warren K. Lewis of the Massachusetts Institute of Technology applied precise mathematical formulations to Little's concept of "unit operations," thus allowing for greater sophistication in research as well as plant design and construction. In 1926 six of DuPont's 10 chemical engineers were MIT graduates. Research director Charles M. A. Stine welcomed their contributions to the company's research ventures.

firm, B.A.S.F. (Badische Analin & Soda-Fabrik), was interested in joining with DuPont to start a synthetic ammonia plant in America. It was understood that B.A.S.F. would share some dyestuffs know-how once the ammonia plant was operational. But when the Germans learned of DuPont's participation in the Chemical Foundation, and of the company's lobbying in Washington to extend the wartime embargo on German dyes, their enthusiasm for a deal waned. Discovering the German firms' deliberate duplicity in filing misleading dye patents in America helped Pierre's younger brother Irénée, who had become president of DuPont in May, justify DuPont's efforts to continue the embargo on imported dyes. With some low-profile help from a newly appointed U.S. senator from Delaware, T. Coleman du Pont, those efforts bore fruit in the Fordney-McCumber Tariff of 1922. Embargos and tariffs would help protect the fledgling domestic industry, but they would not invent any dyes. That difficult task still confronted DuPont's researchers, whose problems were again compounded when the recession struck in 1920.

Nationally, dye production fell by 45 percent from 1920 to 1921. Employment in the Deepwater dye works dropped from 3,500 early in 1920 to 900 by 1921. The staff of 565 at Jackson Lab was cut to 217 in just a few months. All lab assistants were laid off, so demoralized and distracted chemists washed their own glassware and waited for the next discouraging development. Economic uncertainty also aggravated the professional tensions between chemists doing research and chemical engineers trying to solve mass-production problems, prompting the Chemical Department's C. Chester Ahlum, who had been among the team visiting Levinstein, Ltd., in England, to write a memo reminding his Jackson Lab chemists of the need for cooperation with the engineers.[5]

For DuPont and many other companies, the decade that eventually would be tagged "the Roaring '20s" was off to a bleak start. Millions of jobs in the textile, printing, leather and related industries depended on the successful unlocking of the many secrets to making synthetic dyes, and now the keys seemed farther out of reach than ever. Even companies that did not depend greatly on dyes were so short of red ink that bookkeepers had to change the color they used to mark business losses. Having failed to benefit from German patents or strike a useful partnership with a German chemical firm, Irénée and the Executive and Finance Committees finally agreed in July 1920 on a high-risk gamble: DuPont would try to recruit German dye specialists to work at the Jackson Laboratory.

Dyestuffs know-how was so closely tied to explosives and, more ominously, to the poison gases that had spread new horrors over World War I battlefields that DuPont persuaded the U.S. government to help in its recruitment effort. But Germany was trying to recover its thriving prewar dye trade in part by keeping experts in German firms under contractual arrangements it considered legally binding. Ten German specialists eventually broke their contracts, crossed the German border and found new careers in America with DuPont. The first two, Joseph Flachslaender and Otto Runge, left the Bayer Company and successfully made their way to the Netherlands. Two of their colleagues, Max Engelmann and Heinrich Jordan, along with a trunk full of technical documents, were detained in the Netherlands under a German arrest warrant. Over the next few months, U.S. military occupation forces in Germany, assisted by State Department officials, succeeded in gaining the release of Engelmann and Jordan, who arrived in New Jersey on July 5.[6] The Germans brought no magic cures for DuPont's complicated dye problems; they had left Europe under great pressure and could carry only a limited amount of information in their heads. Moreover, the four Germans arrived at a difficult time, since many American researchers had been laid off in the recession and the new recruits were being paid $25,000 a year — about five times the salary of most of the American chemists. Still, their help was appreciated by the Jackson Lab chemists, who overcame some initial misgivings and accepted the newcomers as colleagues.

The recession also added urgency to complaints that midlevel managers were voicing about DuPont's highly centralized structure. Salesmen in the central Sales Department, for example, found themselves promoting several brands of paint under different labels, all from DuPont's acquired companies: Harrison Brothers, Bridgeport Wood Finishing, and New England Oil, Paint, and Varnish. Salesmen also were duplicating each other's efforts, confusing retailers in the process. Just as worrisome, problems with quality control in plants that were not yet completely under DuPont supervision drew ridicule from competitors and threatened the integrity of the company name.

In March 1919 DuPont's Executive Committee commissioned a review of the company's disappointing performance in its diversified product lines and a study of production problems such as those reported with paints. But at that moment the Executive Committee itself was in transition as Pierre, preparing to relinquish the DuPont presidency, looked to a younger generation to carry on the postwar expansion plans. In April Lammot du Pont, the only holdover from the older, wartime committee, was made Chair. He was joined on the committee by his several talented executives, including a fast-rising, 31-year-old manager named Walter S. Carpenter Jr. and F. Donaldson Brown, who had taken Raskob's place as treasurer several months earlier when Raskob assumed full-time duties at GM.

The following year the review committee, chaired by the Development Department's Frank S. MacGregor, reported back to the Executive Committee with recommendations for major changes in DuPont's internal structure. MacGregor had acquired firsthand knowledge of the weak points in the company's structure by helping organize paint operations. His committee recommended decentralizing and restructuring the company's operations. Instead of grouping the various product

TOP LEFT: **Henry G. Haskell organized the High Explosive Operating Department for DuPont in the 1890s and eventually became the manager of High Explosives for the entire company.**

TOP CENTER: **Robert Ruliph Morgan Carpenter, older brother of Walter S., was head of the Development Division and part of the company's leadership.**

TOP RIGHT: **Walter S. Carpenter Jr. led the company from 1940 to 1948. With the exception of Antoine Bidermann, who guided the company for three years after the founder's death, Carpenter was the first president of the company unrelated to the du Pont family.**

BOTTOM: **After deciding to go into the paint business after World War I, DuPont began acquiring paint companies across the country, such as Harrison Brothers of Philadelphia (shown here).**

TOP LEFT: **DuPont's investment in General Motors increased the company's presence in Detroit and prompted DuPont to develop dozens of products for the auto industry.**

TOP RIGHT: **Pierre S. du Pont, depicted as president of General Motors in 1922, in New York City's newspaper *The Globe*.**

BOTTOM CENTER: **Irénée du Pont, younger brother of Pierre S., became president of DuPont in 1919. In 1921 he created the departmental organization of the company that would be its central organizing structure until the 1990s.**

BOTTOM RIGHT: **Alfred P. Sloan Jr. succeeded Pierre S. du Pont as president of General Motors in 1923.**

lines around centralized functions such as purchasing, manufacturing and sales, those functions would be grouped instead around the products. For example, a single, central sales department previously had promoted multiple products; the plan called for each broad product category to have its own sales force specially oriented toward clearly identified markets.

President Irénée du Pont initially resisted the subcommittee's recommendations for change. Irénée, Pierre and other top executives had worked hard for 10 years to attract and keep talented managers at DuPont. They feared that the expertise concentrated in the company's centralized functions might be diluted by spreading those functions out to serve each product line under a department manager. Though Executive Committee members Brown, William Spruance and Frederick Pickard backed the review committee's recommendations, Chairman Lammot, Walter S. Carpenter Jr. and J. B. D. Edge opposed them, as did Irénée. The decentralization plan languished during 1920 and most of 1921.[7]

Meanwhile, DuPont's leadership was learning valuable lessons from General Motors and its new president, Pierre S. du Pont, who had taken over from William Durant in November 1920. In contrast to DuPont's excess of central control, General Motors' several divisions under Durant had operated as virtually autonomous entities. Lacking coordination and direction, they had simply plowed ahead on their own. When the automobile market collapsed in the 1920 recession, its independent divisions, like broken spigots, kept pouring out new cars. By October GM had exceeded by 40 percent the total inventory limit it had imposed on its combined divisions just five months earlier. By November, sales had dropped 75 percent from early summer. The Olds, Oakland and Chevrolet divisions were completely shut down, while the more successful Buick and Cadillac divisions had greatly reduced production. In the flush of wartime expansion, DuPont's Finance Committee had responded to John Raskob's entreaties for additional DuPont investment in GM. By the time DuPont made its last purchase of nearly $5 million worth in August 1919, the company's holdings totaled $50 million — a controlling interest of nearly 30 percent of GM's common stock. When Durant was forced to resign in November, Pierre put aside his long-sought wish to cultivate the gardens at his Longwood estate in Pennsylvania. He stepped into Durant's place to help save DuPont's significant investment.

General Motors was in bad shape, economically and physically. Pierre listened carefully to the rehabilitation plan offered by Alfred P. Sloan Jr., a 45-year-old electrical engineer, GM vice president, and fellow Massachusetts Institute of Technology alumnus. The arguments Sloan offered for aligning GM's divisions according to the markets they served — for greater central planning and forecasting, and for using committees to coordinate these central functions with GM's several production divisions — almost paralleled the MacGregor committee's proposals for improving relations between DuPont's central headquarters and its increasingly diverse production operations.

GM's Board of Directors ratified Sloan's reorganization plan, which went into effect on December 29, 1920. By September 1921 the example of General Motors and the pressure of the recession prompted Irénée and some members of DuPont's Executive Committee to reconsider their opposition to restructuring their own company. Weighing the risks of change against the benefits of adhering to the status quo, they decided to proceed with restructuring. They divided DuPont's manufacturing operations into five main product departments — Explosives, Cellulose Products, Pyralin (plastic), Paints and Dyestuffs — and placed them under five general managers who had carried

Compliments of N.Y. Globe.

DUPONT

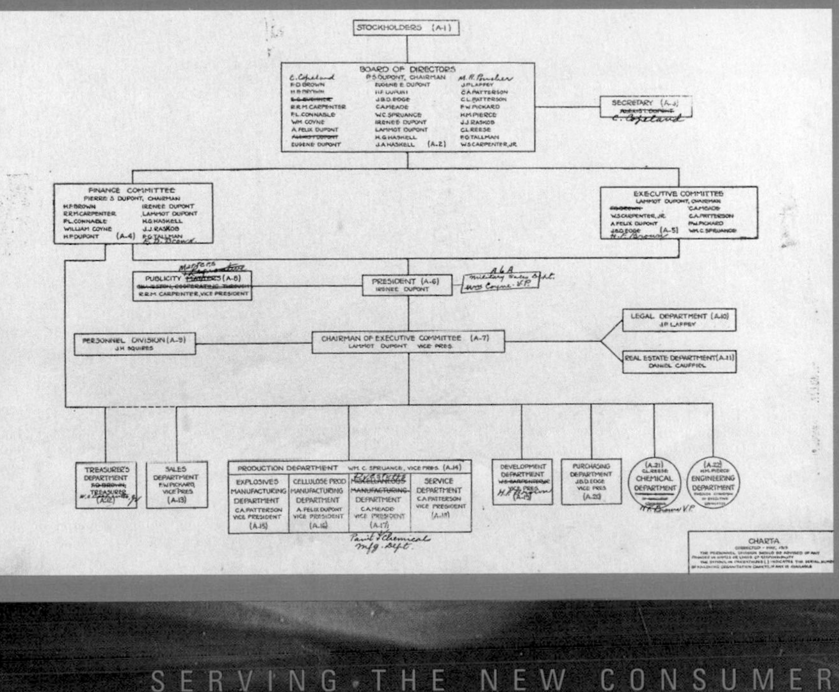

day-to-day responsibility for purchasing, production and sales. Advertising, development, engineering and legal services were retained as central operations rather than being delegated to department-level management. Research and development continued to be largely centralized in the Chemical Department's Experimental Station laboratory, but the 1921 reorganization also affirmed the value of the product-oriented research that had been going on for years in DuPont's industrial departments. Working out the optimum relationship between centralized and departmental research activities continued to be a challenge for DuPont's research directors as the company's needs and the U.S. economy changed over time.

The DuPont-GM relationship quickly became a two-way pipeline for managerial information and talent. Several DuPont executives, such as J. Amory Haskell, F. Donaldson Brown, and John Lee Pratt, a young engineer in the Motor Development Section of DuPont's Development Department, found careers at GM. DuPont's Engineering Department carried out numerous construction and design projects for the auto company, including a large Cadillac plant in Detroit and 1,500 workers' houses in Flint and Pontiac, Michigan. But at the end of 1921 it was not at all clear that the GM-DuPont relationship would continue to be so mutually profitable. While both companies enacted sweeping reorganization plans, DuPont struggled to bring new products to market and anxiously waited for America's dormant consumer market to wake up.

By 1922 there were signs of new vigor in the U.S. economy. From a recession low of $7.5 million in 1921, DuPont's earnings nearly doubled in 1922, jumped to $21 million in 1923, and reached $43 million in 1926. More than half that sum represented dividends from DuPont's GM investment, which continued to provide roughly half of all DuPont's earnings through 1929. In that year DuPont posted record earnings of $82 million.

BACKGROUND: This was the DuPont Advertising Department office in Wilmington during World War I.

OPPOSITE TOP INSET: As company holdings increased, the support staff in Wilmington grew as well.

OPPOSITE BOTTOM INSET: F. Donaldson Brown served on both the Executive and Finance committees at DuPont, in addition to playing a key role in helping Pierre S. du Pont and Alfred P. Sloan Jr. rescue the floundering General Motors in 1920.

LEFT: Pressed by internal troubles and a national economic recession, the Executive Committee proceeded with a major corporate reorganization in 1921. Operations were divided into five main project departments, and functions such as advertising and legal services were retained in a central core.

NOTABLE

> DuPont tried to learn the secrets of German synthetic dyes by visiting plants working under patents confiscated when the First World War began. Chemist Arthur Douglas Chambers was part of a team sent to the Levinstein plant outside Manchester, England, in April 1916. He and J. Amory Haskell returned from England in November 1916 with an agreement to purchase information about synthetic dye chemistry and production. A second DuPont team then spent December 1916 and January 1917 at Levinstein, busily taking notes before having to hurry back to America under a new threat from German U-boats.

Chambers's work at DuPont had not always been so serious. On October 9, 1909, he participated in a "one-day, continuous vaudeville" show staged by and for DuPont Explosives Department personnel at a meeting in Atlantic City, New Jersey. Chambers sang with "The Stability Quintet" and recited his "original gem, All Bachelors Are Shiftless," with J. Amory Haskell's brother, Henry G. Haskell, rendering a "Honk! Honk! accompaniment" with his automobile. In 1944 the Deepwater dye works was named the Chambers Works in his honor.

> DuPont recruited several German dye specialists after the war for its research and development effort at the Jackson Lab. In February 1921, Richard Sylvester of DuPont security and Elmer K. Bolton, the genial, red-headed director of the Chemical Department's Organic Division, strode purposefully toward the Dutch ship *Ryndam*, docked at Hoboken, New Jersey, to greet new arrivals Joseph Flachslaender and Otto Runge. "Bolton," Sylvester said confidently, "we are going over there and get these Germans." When all the passengers had disembarked, however, Flachslaender and Runge could not be found. There were outstanding German warrants for their arrest, and the *Ryndam*'s captain would not release them. Sylvester marched up the gangplank and flashed a set of impressive-looking credentials at the captain: "Major Richard Sylvester, Honorary President, International Association of Chiefs of Police." But the captain was not impressed. Sylvester and Bolton rushed to the nearest telephone to call for help from Wilmington. After several phone calls between Wilmington, Washington, D.C., and authorities at Ellis Island, Flachslaender and Runge were allowed to begin new lives as DuPont chemists. Eventually, eight more German dye specialists joined DuPont's research teams.

> In 1919, DuPont adopted an innovative Group Life Insurance Plan with free coverage of $1,000 to $1,500 for practically all employees who had served for six months or more.

> The "line and staff" organizational structure that Alfred P. Sloan Jr. and Pierre du Pont introduced at General Motors, and that Pierre and Irénée du Pont brought to DuPont, became the model for U.S. industries for the next half-century. This model, adapted from the military, called for a centralized "staff" of executives who were freed from day-to-day management responsibilities in order to focus on setting broader goals for the company. Meanwhile, semi-autonomous "line" managers in the company's many departments or divisions took responsibility for daily production. Regular meetings between line and staff managers helped maintain a balance between centralized control and departmental autonomy.

> The Nobel Group joined with three other British chemical companies in 1926 to form Imperial Chemicals (ICI) to better compete with large chemical consortiums such as I. G. Farben. DuPont immediately reached a patent agreement with the new firm and in 1929, signed expanded patent and process sharing agreements.

> Nitrogen gas is part of the air we breathe and the soil we walk on. It is a vital component of living matter as well as a key ingredient in explosives, certain plastics and other organic materials. Before it can be used by living organisms, however, nitrogen must first be drawn from the air by being "fixed," or combined with hydrogen or oxygen. Certain bacteria in the soil are able to fix nitrogen that plants can then absorb and

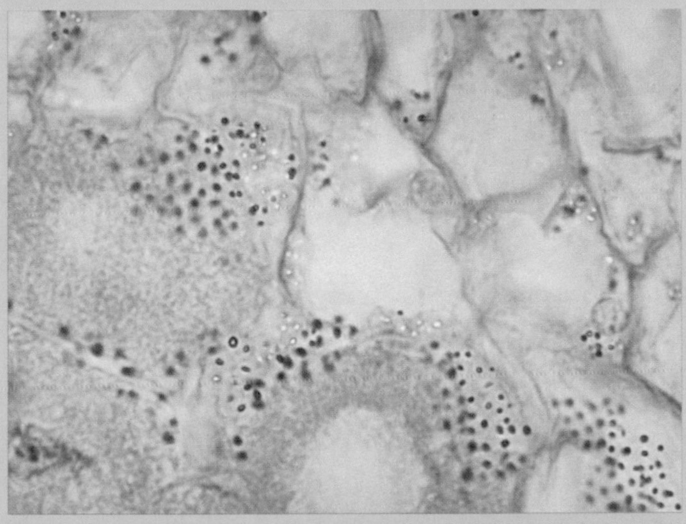

pass along the food chain. Large amounts of nitrogen are then returned to the soil as ammonia (nitrogen combined with hydrogen) in animal wastes that bacteria transform into usable nitrogen for plants. Large potassium and sodium nitrate fields in India and Chile were derived largely from bird droppings and offered "fixed nitrogen" supplies for explosives and fertilizers in the 19th century. But these natural sources were limited and difficult to reach. In the years after 1900, DuPont's Development Department closely watched efforts by European researchers to synthesize ammonia, a valuable source of nitrogen. Fritz Haber and Carl Bosche succeeded with high-pressure ammonia synthesis in Germany in 1910, and George Claude developed an alternative process in France in 1917. DuPont in 1924 joined with a French firm, L'Air Liquide, to open a U.S. plant using the Claude process and in May 1925 started construction of its Belle, West Virginia, plant. The engineering and chemical expertise DuPont acquired at Belle proved valuable in a remarkable variety of products, including methanol antifreeze, urea fertilizers and nylon.

for Better Pictures
DU PONT
NEGATIVE
FILM

By 1929, U.S. manufacturers produced nearly 5 million cars annually. Americans owned 80 percent of the world's automobiles. With an average of one car for every five people, the United States far surpassed the next closest car consumer, Great Britain, which averaged one for every 43 people by the end of the decade. A host of related industries, from road construction to rubber, oil and tourism, drew sustenance from the automobile. Borrowing money shed its older aura of embarrassment as Americans turned to "installment plan" financing as a legitimate way to buy cars, refrigerators and other "big ticket" items. John Raskob, Pierre du Pont's close friend and associate, pioneered automobile financing by creating the General Motors Acceptance Corporation in 1919. By 1925 about 65 percent of new cars were bought on an installment plan, and by the end of the decade consumer lending had become America's 10th largest business. Advertisers were spending $2.5 billion annually to lure consumers into a flourishing market for electrical appliances, telephones, radios, cigarettes, soaps and mouthwashes.

DuPont's chemists helped solve one of the auto industry's most nagging problems in the early 1920s: the length of time it took for the oil-based — and non-lasting — exterior paints of that era to dry. Assembly lines could speed up car production, but no one could figure out how to make paint dry faster. Bottlenecks formed, leaving up to 15,000 cars parked on manufacturers' lots for weeks at a time, waiting for their oil-based coats to dry. DuPont and other companies produced durable and fast-drying nitrocellulose-based paints and lacquers, but they were too viscous to be sprayed. Their use was therefore limited to such applications as dipping brush handles and toys. Large-scale industrial application awaited the development of a nitrocellulose paint thick enough to last but thin enough to be sprayed quickly over large surfaces. As it turned out, the solution to the auto paint problem came when researchers at DuPont's Parlin plant in New Jersey saw possibilities, not failure, in an experiment that went bad.

Like the coal tar chemicals that Jackson Lab's chemists were slowly turning into a rainbow of new dyes, cellulose proved to be a wondrously fertile source of useful materials. When treated with certain acids and processed in the right ways, this basic building material of plant life could be transformed into explosives, plastics, artificial leather and silk, or lacquer and paint. When spread out into thin sheets and covered with light-sensitive emulsion, it even became movie film. DuPont had considered going into film manufacturing as early as 1912, but it was not until after World War I that the company took the plunge into full-scale production in its Cellulose Products Department. In the winter of 1920 a large quantity of DuPont film was ruined when sparks of static electricity speckled the film while it was being rolled up. Chemists at DuPont's Redpath film lab soon discovered that a small amount of sodium acetate added to the nitrocellulose helped drain off the static electricity before it could spark, and in July they prepared a large drum of the viscous mixture for a large-scale trial run.

At that critical moment there was a power breakdown at Parlin that took several days to fix. While the plant idled, the covered drum of film solution simmered, neglected under the slow arc of the summer sun. Inside, the sodium acetate quietly churned toward the conclusions that the laws of heat and chemistry decreed. When power was at last restored, workers carried the drum to the film plant for casting. But when they removed the lid, they found that the thick, jelly-like substance they had mixed several days before had become strangely thin and syrupy — like paint. Chemist J. D. Shiels and plant superintendent Edmund Flaherty looked further into the mysterious chemical reaction. They soon confirmed that something remarkable, and patentable, had occurred. The Cellulose Products Department's Redpath film lab had inadvertently discovered a way to make a low-viscosity, high-nitrocellulose paint.

After three years of extensive testing, including prolonged exposure to the sun as well as harsh rubbing with mud, motor oil and road tar, a light blue Duco paint made its

TOP LEFT: **DuPont chemists discovered by accident how to make Duco paint. This battery of Duco mixers made the final product.**

BOTTOM: **Duco paint transformed the auto industry. Unlike previous generations of paints that chipped, rusted and bruised and took several weeks to dry, Duco paint was fast-drying and relatively indestructible.**

TOP RIGHT: **DuPont began to supply Hollywood with cellulose film for movies in the 1920s.**

debut on the assembly line of GM's Oakland Division. The following year, 1924, GM introduced Duco throughout all its divisions. Duco and related wood lacquers also proved a boon to furniture dealers, who could advertise stain- and water-resistant surfaces. Consumers could now relax about spills, bumps and bangs.

Marketing "brush" Duco for home use, however, proved more difficult than DuPont salesmen had thought. The paint dried so quickly that homeowners in what industry insiders called the "tin can trade" found that they could not "work" it like the oil paints they had always used. It dried before they were finished spreading it, leaving an uneven coat. Though salesmen considered launching a campaign to educate consumers about how to brush Duco onto wood and other surfaces, chemists at the Experimental Station instead went to work to find an oil-like resin that could be added to Duco to improve "spreadibility" without detracting from the paint's durability. Their efforts led to DuPont's line of Dulux alkyd resin paints, introduced in the summer of 1929 for both the "tin can" and industrial markets. Dulux Super White became the leading coating for refrigerators in the early 1930s and, in yet another happy coincidence for DuPont, found a ready market in General Motors' Frigidaire Division.

DuPont had first paid attention to artificial leather in 1909, when a federal antitrust suit threatened its dominance of smokeless powder manufacturing and spurred efforts toward diversification. Artificial leather was basically nitrocellulose mixed with castor oil and spread over a mesh fabric to give it strength. Dyeing and embossing provided a leathery appearance. Believing DuPont's research in this area to be behind that of the Fabrikoid Company in Newburgh, New York, Irénée du Pont, then director of the Development Department, persuaded the Executive Committee in 1910 to purchase Fabrikoid. But when Fin Sparre, chief chemist at the Experimental Station, inspected the Newburgh plant, he was unimpressed. Nevertheless, in a moment of remarkable

prescience, Pierre thought the product might prove useful to car makers. DuPont acquired Fabrikoid, upgraded that company's Newburgh facilities, made some improvements in the product, and found a modest niche for it in the book binding, home furnishings, rain wear and automobile businesses. Fabrikoid automobile tops were much cheaper than real leather tops but did not last well outdoors. As a covering for seat cushions, Fabrikoid would separate when stretched regularly. Nevertheless, automobile-related sales boosted Fabrikoid business by 66 percent in 1923, and the versatile product earned modest profits during the remainder of the decade.

DuPont had followed its paint and artificial leather acquisitions in 1904 and 1910 with still another nitrocellulose purchase that yielded profits in the 1920s automobile business. A prolonged strike in 1913 had exasperated management and morale problems at the nation's largest plastics producer, the Arlington Company of New Jersey. Two years later the discouraged owners sold the 30-year-old company to DuPont for a cash payment of $6.7 million, thereby giving DuPont 40 percent of the U.S. plastics market.[8] Arlington's line of stiff, white, Pyralin shirt collars and cuffs had earned solid profits for years, along with its numerous combs, brushes, hairpieces and eyeglass frames. The company also had produced nitrocellulose lacquers and supplied clear, plastic side curtains for early, open-air automobiles. The increasing popularity of the closed car between 1909 and 1920 reduced sales of Pyralin windows, but the clear plastic product reappeared to DuPont's profit in 1929 as the shatter-resistant layer between twin glass sheets of safety windows. In 1925 DuPont added the Viscoloid Company of Leominster, Massachusetts, to its plastics operations and transferred production of all toothbrushes and toilet articles to that site.

DuPont's robust growth in the 1920s was not all connected to the phenomenal growth of the automobile industry. In 1924 DuPont joined with a French firm, L'Air Liquide, to open an ammonia synthesis plant to produce

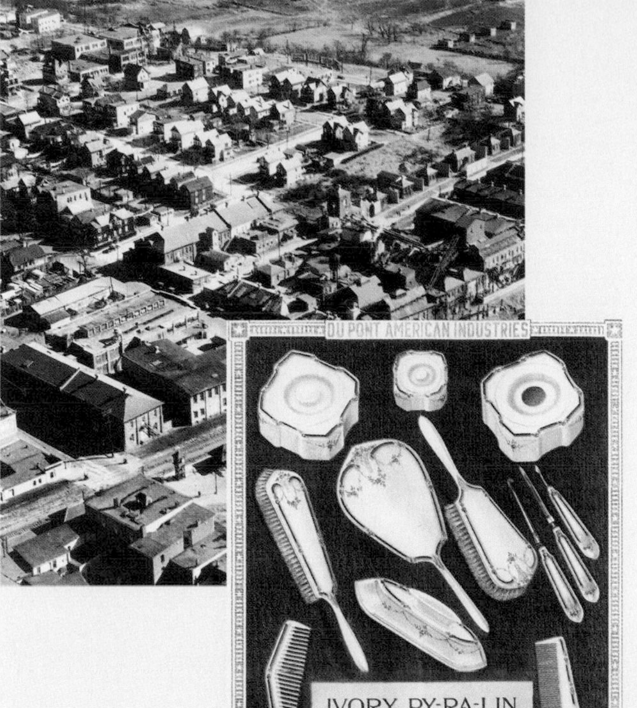

OPPOSITE: **The manufacture of Duco paint advanced from simple hand-filling operations to a full-sized plant in Flint, Michigan, to serve the auto industry. In addition to its impact on that industry, Duco was advertised for home use, such as in kitchens and on appliances.**

DuPont manufactured accessories for the auto, inside and out, including a Fabrikoid car cushion for Chevrolet.

TOP PAIR: **DuPont bought the Arlington Company of New Jersey, entering the emerging plastics field in 1915. The factory turned out Pyralin shirt collars and cuffs, along with a steady stream of plastic toiletry items such as mirrors, combs and toothbrushes through the 1920s.**

BOTTOM PAIR: **Car tops of the artificial leather Fabrikoid underwent weathering tests at the Experimental Station. "A 'tip-over' didn't hurt this top" reads a 1923 magazine article about Fabrikoid tops.**

Count Chardonnet of France tried to replicate the actions of silkworms to produce artificial silk. He succeeded in making a predecessor to rayon.

MIDDLE: The Buffalo, New York, plant produced rayon beginning in 1920 and cellophane beginning in 1924.

BOTTOM: Chemical engineer Leonard Yerkes managed the Buffalo Fibersilk — later rayon — plant in the 1920s.

OPPOSITE: A 1926 DuPont ad for rayon.

valuable nitrogen-related chemicals for fertilizers, explosives and refrigerants. The next year DuPont engineers started construction of the ammonia plant at Belle, West Virginia. In 1928 DuPont's purchase of Cleveland's venerable Grasselli Chemical Company added 23 heavy chemical, paint pigment and explosives plants to its assets. Meanwhile, Duco, Fabrikoid and Pyralin found profitable markets outside as well as inside the auto industry, and some products, like rayon and cellophane, thrived with little or no help from automobile production. Rayon and cellophane were essentially different forms of the same chemical substance, one made by forcing treated cellulose through small holes into an acid bath, where the strands became synthetic "silk" thread, the other made by adding glycerin to prevent brittleness and spreading it out to dry in thin, transparent sheets. These materials were not new in 1920, but they were not commonly used. It took the ability to see the commercial possibilities of exotic laboratory discoveries, along with creative research, production innovations and a determined marketing campaign, to make "rayon" and "cellophane" household words.

As Coleman had intended when he established DuPont's Development Department back in 1903, the unit kept a sharp eye on long-range opportunities for the company. DuPont was still seven years away from its first move into cellulose fibers when the department, under director Irénée du Pont, sent chemist Fin Sparre to Europe in 1909 to investigate fascinating new products such as artificial silk. Louis Bernigaut, a protégé of Louis Pasteur, had been studying silkworms in southern France in the 1880s. Bernigaut, who bore the title Comte de Chardonnet, observed that the worms took in cellulose at one end in the form of mulberry leaves, then extruded it at the other end in the form of silk thread. By 1891 Chardonnet had succeeded in replicating this process in the lab, and the following year two Englishmen, Charles Cross and Edward Bevan, discovered a way to make a thicker, more pliable thread they called Viscose.[9] Fifteen years later, Jacques

Edwin Brandenberger, a Swiss chemist working in France, tried to find a practical use for the transparent cellulose film he had invented. He went to the Comptoir des Textiles Artificiels for assistance, but in the end he could market his "cellophane" only as an expensive gift wrap.

In 1910 the British textile firm Courtaulds, Ltd. set up a Viscose plant in America. Within two years the company reaped an impressive return of almost 80 percent on its investment. DuPont tried to buy the plant in 1916 as part of its preparation for diversified, postwar production, but the American Viscose Company had been too profitable for Courtaulds to let go. Late in 1919, however, as DuPont's Experimental Station developed models for the company's own artificial silk production, an offer for a joint venture in America came in from the Comptoir des Textiles Artificiels. Leonard A. Yerkes, a chemical engineer and assistant director of DuPont's Development Department, went to France to investigate. His report was favorable, and in April 1920 Yerkes became the first president of the jointly owned DuPont Fibersilk Company. The operation produced its first yarn in May 1921 in a Buffalo, New York, plant that DuPont bought from the Philadelphia Rubber Works Company. Three years later textile manufacturers decided to promote the new synthetic by renaming it "rayon," a combination of "ray" for its luster and "-on" from "cotton." The debt to the silkworm was thereby rhetorically canceled, and rayon stood as a product in its own right. Accordingly, DuPont Fibersilk Company became DuPont Rayon Company in March 1925.

Several months after the Buffalo plant opened, officials from the Comptoir des Textiles Artificiels mentioned that they had another product that might interest DuPont. Early in 1923 William Spruance Jr., a new member of DuPont's Executive Committee, traveled to France and brought samples back to Wilmington. The thin, translucent substance the

Dr. Hale Charch tried 2,500 formulas and made hundreds of tests to develop a lacquer coating that made cellophane moisture proof, leading the way for the product's explosive consumer impact.

OPPOSITE TOP: As the company grew through the 1920s, more and more office staff had to be added to handle the business.

OPPOSITE BOTTOM LEFT: Ernest B. Benger was the Chemical Director of the DuPont Rayon Company in 1925 when he hired Hale Charch to work on the development of moisture-proof cellophane.

OPPOSITE BOTTOM CENTER and RIGHT: The inspection of cellophane at Waynesboro, Virginia.

French called cellophane did not stand up well under ocean travel, nor under the scrutiny of Fin Sparre, now director of the Development Department, who dramatically mashed the yellowed sheets into a crinkled pile on the table. Despite Sparre's reservations, DuPont signed a joint agreement with the French firm in June 1923. The DuPont Cellophane Company began production at the Buffalo plant in 1924, even though company researchers were still trying to find a way to iron the wrinkles out of Brandenberger's invention.

Chain grocery stores, where customers bought merchandise off the shelf rather than being served by clerks, tripled in number in the 1920s, creating a huge market for attractively packaged goods. By the end of the decade A&P's billion-dollar-a-year business amounted to 10 percent of all U.S. retail food sales. But in 1924 cellophane's uses were limited to cookies and chocolates. DuPont's salesmen were kept out of the lucrative self-service, food-wrap trade because cellophane was not moisture-proof. In fact, ordinary wax paper was cheaper, and 98 percent more moisture-proof, than cellophane. A bag made of cellophane could hold water without leaking, yet water vapor could evaporate through its finely porous surface. DuPont had marketed the product as waterproof, which was technically correct but also a bit misleading, for it was indeed waterproof but not moisture-proof. Until that problem could be solved, DuPont would fare no better with cellophane than Jacques Brandenberger had.

One day in 1925 Ernest Benger, chemical director of the DuPont Rayon Company, looked dourly at the resume of William Hale Charch. Just 27 years old, Charch had recently been laid off at GM, where he had worked for two years with engineer Thomas Midgley on anti-knock gasoline additives. Charch had few qualifications besides his recent Ph.D. in chemistry from Ohio State University and a measure of personal charm. Benger was not much interested but yielded to his secretary's entreaty to give the young scientist a chance.

Emulating Midgley's empirical techniques at GM,

That sparkling transparent wrapping is Du Pont Cellophane
This photograph taken through a sheet of Cellophane
Of course, you have seen it countless times in a host of different places. A glistening, transparent sheet that serves to show the product it encloses and keeps it clean and attractive. It is Cellophane and is made by du Pont.
Through it you see what you buy before you buy. Then, too, a wrap of Cellophane tells you a manufacturer desires to have his product at its best. It is a guide to quality.
As you remove this sparkling wrap you realize that the article is reaching you just as it left the makers. The protection of Cellophane guards it from dust and all contamination.
At your grocer's you see a wide variety of quality products displayed in this manner. Luscious fruits, appetizing meats, tempting baked goods and countless other products are displayed in this manner. You can purchase them with confidence.
Du Pont Cellophane Company, Inc., Sales Offices: 2 Park Avenue, New York City; Canadian Agents: Wm. B. Stewart & Sons, Limited, Toronto, Canada.
DU PONT CELLOPHANE
Cellophane
STEEL CUT COFFEE
OLIVES
Sparkling Cellophane—The Individual Gift Wrap
The presents you give on Christmas Day will look their best in a brilliant wrap of red or green Cellophane—a fitting symbol of the holiday season.
This beautiful wrap gives the gift a touch of individuality that always appeals to the recipient.
In the new Dennison household unit there is enough Cellophane for wrapping several gifts. This popular unit sells at fifty cents and can be obtained at your nearest Dennison dealer, or by writing to the Dennison Manufacturing Company, Dept. 200, Framingham, Mass.
DU PONT CELLOPHANE COMPANY, INC.
2 Park Avenue, New York City
Canadian Agents:
Wm. B. Stewart & Sons, Limited, Toronto, Canada
DU PONT CELLOPHANE
Cellophane
THE BOON TO THE CIGAR INDUSTRY
"A FRESH five-cent cigar is a far better smoke than a dried-out one that costs a dollar," says an authority. Manufacturers, realizing this, have long sought to keep their products fresh and smokable. Some time ago the packaging problem became acute, because consumers insisted on seeing their smokes before buying them. The development of moisture-proof Cellophane has made that possible; it displays cigars perfectly and yet keeps them moist, full of aroma, and fresh indefinitely. Now we can choose our favorite shape and shade of cigar with confidence in its condition. What a boon! And how comforting it must be to the manufacturer to know that modern chemistry has provided a suitable transparent wrap that keeps his products as fresh as when they left the factory! A few of the brands obtainable in Cellophane are illustrated here.
LA PALIN
NATURAL BLOOM
Haddon Hall
HAVA-TAMPA
NEW NURICA
192-267

Church tried anything that might make cellophane moisture-proof, including latex rubber, wax and various nitrocellulose substances. With his assistant, Karl Prindle, Church soon focused on a combination of nitrocellulose and wax. The challenge was to apply the super-thin, wax-infused lacquer to cellophane with solvents in such a way that when the solvents evaporated, the wax coating would bind without changing the cellophane's transparency or strength, and without adding a sticky feel. Within a year Church and Prindle had found the right combination of coatings and solvents, only five one-hundred-thousandths of an inch thick and applied to both sides of the cellophane. Eighteen months later DuPont engineers had designed and constructed the necessary manufacturing apparatus.

After filing the patent for its new, moisture-proof cellophane in January 1927, DuPont launched an expensive, three-year advertising campaign in the *Saturday Evening Post*. DuPont salesmen could now boast of a product twice as moisture-proof as wax paper. Long-range investment, creative innovation, and skilled product development brought big rewards, and the company shared them with Hale Charch. In 1929 the novice chemist who just a few years before had nearly been shown the door received one of the largest individual bonuses in the history of the company. Charch himself, in addition to the product he made so useful, had proved to be another calculated risk that paid off.

DuPont's willingness to take such risks increased during the 1920s precisely because the company's long shots had paid off. Even the frustrating dyestuffs business pushed the company's development into new areas. At the Jackson Laboratory, Max Engelmann's talents were applied to seed disinfectants and lumber preservatives, while Joseph Flachslaender helped set up the process for making tetraethyl lead gasoline additive. Synthetic camphor, a key ingredient in preparing nitrocellulose acetate plastics, was produced at Deepwater after the war, as were a variety of "rubber accelerators" for speeding up the curing or vulcanization of rubber. Even dyes eventually rewarded the effort and risk their development had demanded. By 1923 DuPont at least stopped losing money on them, and in 1928 the Dyestuffs Department reported a modest profit. Yet DuPont's profit from dyestuffs went far beyond any simple adding-up of dollars. The experience and expertise the company acquired in its protracted dye venture benefited many other projects and product lines, adding value in ways that could not always be traced with an accountant's pen. DuPont thrived not only because its products were so diverse but, more fundamentally, because their underlying chemistry was so similar.

No one saw or expressed this core truth about DuPont more ably or frequently than chemist Charles M. A. Stine, who had overseen dye research in the Chemical Department's Organic Division and succeeded Charles Reese as director of the Chemical Department in May 1924. Stine espoused his vision of the "chemical kinship" among DuPont's varied products in a way that energized DuPont's departments and gave his listeners a renewed appreciation of their work. "The logical development of the company's interests," he told guests at an Explosives Department banquet in Wilmington in the winter of 1925, "has followed lines suggested by the chemistry of those products."[10]

Stine's vision formed the basis for a new era of research at DuPont shortly after Lammot succeeded Irénée as company president in 1926. Stine called it "fundamental" research, not because it was "pure" or removed from immediate practical application, but because it was likely to prove useful somehow, somewhere, in the extended family network of DuPont products. As that network grew and prospered in the 1920s, the discovery of new chemical relations and new products seemed more and more likely. DuPont's penchant for the calculated risk and the long-term gamble soon drew the company deeper into the mysteries of chemical discovery.

LEFT: Hale Charch's development of moisture-proof cellophane was so important, he reenacted a test for this photo. He is touching a bag that held water for weeks, while control bags made of untreated film showed evaporation losses within a few days.

CENTER: Before it was made moisture-proof in 1927, cellophane was advertised as a colorful decorative wrap. Cellophane was used mainly to wrap cigars, candies and other novelties

RIGHT: The leadership of chemists Charles Reese (left) and Charles Stine helped solidify DuPont's reputation as a science company in the 1920s.

CHAPTER 5 DISCOVERY

TOP: **Lammot du Pont (1880–1952), younger brother of Pierre and Irénée, became president of DuPont in 1927.**

BACKGROUND: **Lammot, who was known around Wilmington for riding his bike to work, took a leisurely ride with his wife Bertha in Florida in the 1920s.**

BELOW: **Dr. Charles Stine, appointed Chemical Director in 1924, persuaded the Board of Directors to set aside money for fundamental research. His lab launched DuPont in new directions in the 1930s.**

To Charles Stine, chemical discovery was like rounding up all the strangers at a family reunion. He knew they would be related; he just had to find out how. DuPont's many products were the relatives he knew; in the larger world of organic chemistry were the strangers he had yet to meet. Stine's idea was to organize chemical reunions, bring in some first-rate genealogists and start making connections. The results, he was sure, would be promising and profitable products.

Stine made it all sound easy, "a quest after facts about the properties and behavior of materials, without regard to a specific application of the facts discovered."[1] But when he presented his proposal for a fundamental research program at DuPont before Irénée and the Executive Committee in December 1926, he was asked some probing questions: What sort of chemical groupings would he investigate? Shouldn't such work be done by universities instead of industries like DuPont? How would the program be staffed? How much would it cost? Lacking Stine's vision and confidence, the committee needed the reassurance of greater detail.

On March 31, 1927, Stine provided the details to the group and to the company's new president, Lammot du Pont, who had succeeded Irénée when he retired in January. Like his older brothers, Lammot had learned his science at the Massachusetts Institute of Technology and his business at DuPont. But as head of the company's Miscellaneous Manufacturing Department, Lammot also had gained a keen understanding of the "chemical kinship" among DuPont products. His outlook was an interesting blend of visionary research and tough-minded, even blunt, practicality. Lammot du Pont, like Stine, was convinced that DuPont could have it both ways, and that vision and practicality need not conflict.

Stine asked for $200,000 for the first year, most of it to go toward new research labs at the Experimental Station. He insisted that funding be sustained despite the ups and downs of the business cycle — "patient money," he called it. Because basic research paid off only if it was sustained over the long run, Stine argued that it should be shielded from management trimming during hard times. He proposed a program to explore five broad areas of chemical research: catalysis, colloids, polymerization, chemical engineering and chemical synthesis. Stine maintained that universities could not be relied on to conduct the kind of research most likely to prove useful to industry, but he did propose hiring from academia "one first class research man" to direct each line of work. In early April the Executive and Finance Committees approved Stine's plan, placing DuPont in the company of firms like Eastman Kodak, Bell Telephone and General Electric, which already had established their own basic research programs. Stine had hoped to announce the news in the trade journal *Industrial and Engineering Chemistry*, but Lammot curbed Stine's enthusiasm. "Saw the wood," said DuPont's laconic new chief, "and let publicity take care of itself."[2]

Like the Executive Committee, Harvard's Wallace Hume Carothers was at first skeptical about Stine's program. Carothers was one of the select group of researchers whom Stine and his Experimental Station colleagues, Arthur Tanberg and Hamilton Bradshaw, tried to recruit. Carothers had been born, raised and educated in the Midwest, receiving his Ph.D. in chemistry from the University of Illinois in 1924. Recognized by his professors and peers as a research genius, the 28-year-old taught at Illinois for a year before taking a position at Harvard in 1926. Harvard offered the prospect of

TOP: **Dr. Wallace Carothers was reluctant to leave Harvard, but Dr. Charles Stine, director of the Chemical Department, persuaded him that he would lead a team at DuPont that would be given financial support and facilities unmatched in the academic world.**

BOTTOM: **Carothers arrived at the Experimental Station in 1928 to work in an environment designed to isolate his team from the distractions of immediate operating problems. The men were free to choose their own projects.**

BACKGROUND: **DuPont emphasized the image of the chemist in its advertising and publicity in the 1920s and 1930s. This idealized scientist looks through his microscope on the verge of new discoveries.**

new research facilities and a prestigious career, but as a professor he would have to teach, and the introverted Carothers found lecturing an ordeal. DuPont offered freedom from the classroom and a higher salary, but Carothers did not believe he would be truly free to pursue his own research interests.

After visiting Wilmington and the Experimental Station in late September 1927, Carothers rejected DuPont's offer. Tanberg wrote a follow-up letter encouraging him to reconsider, and Carothers's lengthy reply closed with a surprising personal revelation: "I suffer from neurotic spells of diminished capacity which might constitute a much more serious handicap there than here."[3] Carothers, it turned out, suffered from acute depression. Within a week, however, Bradshaw was in Cambridge, reassuring the uncertain professor once more that at the Experimental Station he would be completely free of any pressure for commercial success. The mild-mannered Bradshaw won Carothers over. After finishing out the semester at Harvard, Carothers reported for work on February 6, 1928, at the Experimental Station's new lab in Building 228, dubbed "Purity Hall" by company wags for the "pure" science conducted there. He would head DuPont's polymer research program.

The term "polymer," or "many parts," refers to the large molecules composed of linked smaller units called "monomers" that make up such organic substances as rubber, plastics and cellulose. In Carothers's day a theoretical debate raged about the chemical nature of polymers. Were they large molecules, as German chemist Hermann Staudinger hypothesized in 1920, each one constructed like a chain, with small units repeatedly linked end-to-end by complementary bonds? Or were they large aggregates of small molecules simply massed and held together by some still unknown chemical force? The latter was the accepted theory through the 1920s. After all, as one opponent of Staudinger's view said in 1926, "We are shocked like zoologists would be if they were told that somewhere in Africa an elephant was found which was 1500 feet long and 300 feet high."[4]

Carothers agreed with Staudinger and went to Wilmington in part to prove him right. But instead of trying to break complex polymers down into their component parts, he planned to approach the problem from the other end and put polymers together — synthesize them — as linked, chemical chains of unprecedented length and weight. The creation of such superpolymers would put to rest once and for all the older view. After all, a hopelessly tangled knot of circular shapes might or might not be a necklace. One could argue either side indefinitely. But if a necklace could be built up, link by link, and then massed together into the same "hopeless" knot, there could be no question of its fundamental composition, be it three feet, three yards, or 300 yards long.

DuPont's interest in synthesizing polymers dated back to the mid-1920s, when Willis Harrington, the tough-minded general manager of DuPont's Dyestuffs Department, decided there was at last enough money in the department's research budget for a long-shot project. When he asked Elmer K. Bolton, research director at the Jackson Lab, for ideas, Bolton recalled an incident from his student days in Germany. He had mistakenly understood the tires on Kaiser Wilhelm II's automobile to be made of synthetic rubber — another triumph of German chemistry, he had thought at the time. German chemists had indeed synthesized a rubbery polymer but it performed so poorly that production ceased after the First World War.[5] Producing synthetic rubber had been a goal for industrial chemists ever since, and with Harrington's approval, Bolton proposed making it by polymerizing a carbon-hydrogen gas, butadiene, that he would derive from acetylene, a common but highly flammable gas also composed of carbon and hydrogen. He began to rethink this plan, however, after attending a 1925 conference in which Notre Dame chemistry professor and acetylene expert Father Julius A. Nieuwland raised the possibility of polymerizing acetylene itself. DuPont chemist Arnold Collins took up the acetylene work at Jackson

THE DU PONT
MAGAZINE

EFFECTIVE AT ONCE
WE ARE ADOPTING THE NAME

NEOPRENE

to describe our chloroprene rubber which has previously been sold under the trade mark "DuPrene."

The product itself has not been changed in any way and is exactly the same as that previously sold under the trade mark "DuPrene."

In addition to providing a short generic name for polymerized chloroprene, NEOPRENE can be used to describe the products made from it which display its many distinctive characteristics.

E. I. DU PONT DE NEMOURS & CO., INC.

Rubber Chemicals Division

DECEMBER 16, 1936

A-7254△ PRINTED IN U. S. A.

A story of

MAN-MADE RUBBER

Lab. He failed to produce synthetic rubber but did discover an oil coating that could protect metals from corroding in harsh environments. With this success to his credit, Collins was assigned to Carothers's polymerization group early in 1930.

Rubber was a good example of the knotted necklace problem in polymer research. Since 1860 chemists had known that the basic molecular building block of rubber was a substance called isoprene. But since no one had been able to link the complex isoprene molecules together to make rubber, it remained an open question whether rubber consisted of a chain of regularly linked isoprene molecules, joined end-to-end to make one giant isoprene molecule, or simply a mass of unlinked isoprene molecules congealed into rubber by a mysterious force. Under Carothers's guidance, Collins tried to make a molecular "rubber" chain not from isoprene but from its cousin, butadiene, using acetylene compounds and the catalyst identified by Father Nieuwland.

On April 17, 1930, Collins pried loose a small, solid mass from the bottom of a flask. He turned the pliant substance in his fingers, squeezed it, then tossed it on the floor. As he thought it might, the blob bounced briefly then quivered to a halt. Collins and colleague Julian Hill showed the substance to Carothers, who named it chloroprene because it contained chlorine. Some of his researchers, however, wryly called it just another "popcorn polymer" hatched overnight in the lab and probably of little ultimate value.[6]

Two months later Stine invited Oliver M. Hayden, a rubber specialist from the Dyestuffs Department, recently renamed the Organic Chemicals Department but commonly called "Orchem," to visit his office in the headquarters building. As they drove in his Studebaker Roadster, Stine fished an amber-colored, jelly-like substance from his pocket and handed it to Hayden. Hayden pulled, squeezed and smelled it. It "passed the tooth test," he said, attesting to its obvious rubbery qualities. But Stine already knew that. What he wanted from Hayden and Orchem was a tougher test for

OPPOSITE LEFT: **In one remarkable month — April 1930 — members of the Carothers team discovered the chemical forerunners of neoprene and nylon.**

OPPOSITE RIGHT: **The name change from "DuPrene" to neoprene reflected, in part, the concern that DuPont could not control its use. Neoprene, a generic name coined by DuPont, conveyed the idea that it was an ingredient, not a final product.**

LEFT: **Years after the discovery, Dr. Arnold Collins demonstrated the process for making neoprene in a DuPont lab.**

Arnold Collins's chloroprene: was it good for anything?

Orchem's small rubber mill quickly reduced the Purity Lab sample to useless crumbs. It was not an auspicious start. But the experienced Hayden added wood resin and magnesia and soon had a more workable substance which he called neoprene, or "new rubber." The next challenge was to produce the neoprene from scratch at a pilot plant at Deepwater. This involved handling divinyl acetylene, or DVA, a highly explosive byproduct of acetylene, as well as overcoming numerous other technical difficulties. For example, Collins's neoprene tended to continue polymerizing spontaneously in storage, knotting up its long molecular chains so badly that the material toughened and became worthless. Orchem researchers had to find a way to stop, as well as to start, the process.

Neoprene also emitted the corrosive and toxic gas hydrogen chloride, which Hayden discovered one day when he opened a desk drawer in which he had left some flattened sheets of neoprene on top of his papers. When he picked up the papers they crumbled to pieces in his fingers. Last but not least, the stuff stank. Hayden described the odor as "a mixture of the essences of onion and garlic with a substantial amount of skunk oil added, and then sprinkled with some turpentine."[7] With the notable exception of bad odor, these obstacles to successful commercialization of neoprene were removed, and at a November 2, 1931, meeting of the American Chemical Society, DuPont proudly announced that its Deepwater plant would soon start producing a new synthetic rubber trademarked DuPrene.

When the April 17, 1930, stir over Collins's bouncing polymer had settled in the lab at the Experimental Station, Julian Hill returned to his own bench that day to carry on his work polymerizing esters, organic compounds made from alcohols and acids. Hill was trying to make a long-chain polyester but had hit an upper limit because of a water byproduct that broke the links at a certain point. He and Carothers hoped that if the reaction could be carried out in a high-temperature distillery, or "still," the water could be boiled off and the polyester chain would keep multiplying, yielding a high-strength "superpolyester." The previous day's work with a special still obtained from Johns Hopkins University had been promising, and today, in an astonishing coincidence, Hill matched his lab mate by producing the first polyester superpolymer. Here, in the same week, in the same building, were two inspired proofs of Staudinger's theories about polymers. Carothers, Collins and Hill eagerly prepared their results for publication.

For two more weeks Hill worked at his molecular still, making more of the strong polyester. Then, on April 28, out of some intuitive curiosity, he lowered a glass rod into one of the flasks he had just removed from the still, gently stirred the molten substance on the bottom, and slowly lifted the rod. To his surprise a thin strand of syrupy polyester followed. He pulled some more. The strand lengthened. It was not brittle or fragile. It did not harden or break. In fact, something remarkable happened as Hill drew the filament out farther into the cooling air. The thin, flexible thread became even stronger as the molecules lined up end-to-end in an orderly polymeric chain. "You can just feel the hydrogen bonds take hold," Hill later stated in describing the phenomenon of "cold drawing."[8]

Despite its combination of tensile strength and elasticity, Hill's new "3-16 ester" had serious liabilities as a commercial fiber. It sagged in hot water, melted under ordinary ironing temperatures, and could not stand up to common dry cleaning chemicals. Nevertheless, Hill and Carothers took a sample to Hale Church, inventor of moisture-

OPPOSITE TOP: **Neoprene, invented in 1930, spent the next several months in lab testing before being announced to the public in 1931. The process went on-line at DuPont's Chambers Works in New Jersey.**

OPPOSITE MIDDLE: **This is neoprene emerging from a machine in an early phase of processing.**

OPPOSITE BOTTOM: **DuPont chemist Julian Hill reenacts his April 1930 cold drawing of the first superpolymer fiber.**

Wallace Carothers displays a sample of neoprene, the first commercially successful synthetic rubber.

proof cellophane, at DuPont's rayon plant in Buffalo, New York, for further appraisal.[9] Church was intrigued, as was General Manager Leonard Yerkes. "Well, gentlemen," Yerkes finally said to the Purity Hall researchers, "this is a very interesting observation. Let me know every year or two how you're coming along."[10] Carothers and Hill returned to Wilmington.

Back at the Experimental Station, Elmer Bolton had replaced Charles Stine as director of research and Stine, now a vice president, had joined the Executive Committee as research advisor. Bolton's view of fundamental research was not much different from Stine's, but the overall context for the program was rapidly changing. The nation was in the early stages of an economic depression following the disastrous stock market crash in late October 1929. DuPont's 1930 net income had dropped 30 percent below the 1929 level. With sales volume down 18 percent, and in the face of what Lammot described as a "general dullness of business," the company laid off 4,000 of its 35,000 employees in 1930.

Nevertheless, the Executive Committee honored its commitment to shelter the basic research program from economic downswings. The company also purchased the Roessler and Hasslacher Chemical Company and started a joint venture, the Kinetic Chemical Company, with General Motors to produce Freon® 12, a new chlorofluorocarbon refrigerant that Thomas Midgley had invented. Consumers were wary of older refrigerants such as the toxic gas sulfur dioxide, but DuPont technicians subjected Freon® to extensive testing in order to show that it would neither ignite nor prove toxic under most conditions. This helped to reassure the public and boost refrigerator sales throughout the decade, despite the depressed economy. Few people, including DuPont's executives, thought the Great Depression would last as long as it did. The economy would soon recover, they believed, as it always had in the past.

Late in 1931 Carothers, who had warned DuPont about his mental depression when he was first hired, showed Julian Hill a cyanide capsule he kept fastened to his watch chain. Hill thought it "gruesome," especially in light of Carothers's ready recitation of the names of famous chemists who had committed suicide, but wrote the incident off as mere bravado. Carothers's outlook was indeed darkening. After returning from Buffalo he informed a friend that he and his team had made not only a synthetic rubber but also a synthetic silk. "If these two things can be nailed down," he said wearily, "that will be enough for one lifetime."[11] Carothers's morbid gloom was palpable, but he was so brilliant, and the polymer research at the Experimental Station was going so well, that it was hard for his companions to recognize his melancholy as deep or even real.

On a national level, the outlook also was dark as the Hoover administration struggled unsuccessfully to balance its commitment to government thrift with its grudging acceptance of the need to use federal funds to help shore up the nation's collapsing private businesses. The term "depression" had sometimes been used to describe previous economic downturns, but it was Herbert Hoover's frequent use of this word that associated it with truly severe slumps. Ironically, Hoover's intentions were the opposite. He had wished to boost public spirits by using "depression" instead of the customary terms "crisis" and "panic," which he thought sounded too alarming.[12]

In 1931 the long-awaited recovery had not yet arrived. "We are passing through no ordinary depression," Lammot acknowledged.[13] Major product lines, such as explosives, paints, pyralin sheeting, ammonia and other heavy chemicals, usually accounted for 60 percent of DuPont's sales volume, but their sales were off by more than 10 percent following the slump in basic industries like mining, automobiles and other durable goods. The construction business, which used many DuPont products, declined 78 percent between 1929 and 1933. Additional layoffs of 2,000 employees proved necessary in 1931, though like many enlightened companies at the time, DuPont started reducing work hours and salaries in all its operations to help preserve jobs and retain skilled workers.

Despite deepening economic woes, DuPont enjoyed moments of optimism in 1931. Declining demand for DuPont's heavy-industry products was nearly offset by increasing sales of consumer-oriented products like cellophane, rayon and dyes. There was also the DuPrene discovery that validated the company's support of Stine's fundamental research program. The Organic Chemicals Department, which had successfully developed chloroprene and would now produce it, added greatly to its dye research and production capabilities when DuPont purchased the Newport Chemical Company in August. That same month the company supplemented its zinc- and barium-based (lithopone) white pigment for indoor paint by forming, from its Krebs Pigment and Chemical Company acquisition, the Krebs Pigment and Color Corporation, organized under DuPont's Grasselli Chemicals subsidiary, to make titanium dioxide (white) pigments that produced a better coating on interior surfaces.

There were few such high notes for DuPont, or for the nation, in 1932. By the end of that year, U.S. industrial production had dwindled to 50 percent of mid-1929 levels. Nationally, wages and salaries had fallen 40 percent since 1929, and 12 million workers — 25 percent of the labor force — were idled. Corporate profits nationwide fell 90 percent between 1929 and 1933. DuPont's 1932 sales volume dropped 27 percent below the already depressed 1931 level. The company reduced salaries 10 percent in 1932 and laid off another 1,000 employees, leaving 80 percent of the 1929 total of 35,000. DuPont's net earnings in 1932 shrank to $26 million, less than half of the previous year's total and only one-third of 1929's net profit. Per-share earnings of DuPont's stock, which had been $7.09 in 1929, fell to just $1.82 in 1932.

Franklin Delano Roosevelt's inaugural message on March 4, 1933, "We have nothing to fear but … fear itself," struck a note of hopeful reassurance for millions of desperate Americans. But DuPont's executives believed they had a great deal to fear in Roosevelt. Lammot had been greatly alarmed by the Hoover-sponsored Revenue Act of 1932, which imposed a 33 percent tax increase on corporate profits, the largest peacetime increase in U.S. history. Though he expressed early, cautious support — more caution than support, really — for what he called FDR's "adventurous attack" on the Depression, Lammot quickly became distressed by federal spending programs that, as he saw it, "shock our judgment and defy common sense."[14] He particularly disliked government recommendations to curtail industrial research on grounds that more efficient manufacturing methods raised unemployment. Lammot reassured Bolton that "it is more important to carry out research than to pay dividends."[15] But even this defiant show of support could not insulate DuPont's fundamental research program from the effects of the Depression. Bolton understood from his days with Willis Harrington in Dyestuffs that support was a two-way street, and that DuPont's faith in blue-sky research eventually had to be grounded in profitable results. Stine's vision had been inspiring, but it was Bolton's balanced management approach that made it possible for DuPont to keep what Stine later called an "unwavering faith in research."[16]

Figuratively speaking, hard times forced Elmer Bolton to turn Stine's open-door chemical family reunions into invitation-only affairs. In Carothers's lab that meant greater pressure to come up with a marketable superpolymer fiber to replace rayon, whose surface sheen had suddenly gone out of fashion with clothing designers and consumers. Since 1930 Carothers had been busy with his own publications as well as with numerous professional responsibilities, such as reviewing books and journal articles. Additionally, he estimated that he and his research staff had filed about 60 U.S. and foreign patent applications. But in none of this, as Bolton was fond of saying, could he yet hear "the tinkle of the cash register."[17] "Wallace," Bolton suggested, "if you could just get something with better properties, you would have a new type of fiber."[18] Carothers fretted over the pressure for commercial

TOP: Elmer K. Bolton became director of the Chemical Department in mid-1930, functioning as chief spokesman for DuPont research and development and coordinating the company's entire research program.

BOTTOM: Freon®, a refrigerant invented by Thomas Midgley at GM, was produced in a joint venture with DuPont. Advertising touted the advantages of air conditioned rail travel.

results, but he also knew how the bills were paid and soon reported back to Bolton with some promising ideas.

Esters are formed from acids and alcohols. Amides, another group of organic substances, are formed from acids and ammonia derivatives called amines. Like esters, amides can be polymerized to form long and strong carbon chains, but unlike esters, they have high melting points that would make processing difficult. Carothers's idea was to make a superpolymer fiber with a melting point high enough to stand up to normal fabric use yet low enough to be processed and spun in large quantities. But first he had to figure a way to make a superpolyamide.

On May 23, 1934, Don Coffman, a chemist from the University of Illinois who was assisting Carothers, successfully realized one of his chief's ideas by producing a tough but workable superpolyamide. Carothers and Coffman had proved that it could be done, but the rare and expensive ingredients they used meant that this substance would never leave the lab. The search for the right superpolyamide continued as Carothers and his staff laid out more than 80 possible combinations of acids and amines to explore.

On July 27 chemist Wesley Peterson synthesized a polyamide the team called "polymer 5-10" for the five carbons in its amine portion and the 10 in its acid. Carothers pronounced polymer 5-10 a complete success and expected the substance to proceed immediately toward development, but he was overruled by Bolton, who saw that the "10" portion would have to come from castor oil. It was not the flavor but the anticipated expense of castor oil that soured Bolton on 5-10. As long as its main use was as a foul-tasting medicine, castor oil would remain an inexpensive byproduct of the tropical *Ricinus communis* bean. But in the huge quantities that a successful synthetic fiber would demand, the cost of castor oil would skyrocket. No matter how good a 5-10 fiber might prove to be, the cost of its raw materials would have to be low enough to yield an ultimate profit. So once again Bolton sent Carothers's chemists back to their lab benches.

Seven months of methodical searching ended on February 28, 1935, when Gerard Berchet, whom Carothers had helped recruit from the University of Colorado, produced the polymer 6-6 from the amine hexamethylene diamine and adipic acid. Carothers did not think it was as good as 5-10. Though it stood up well to dry cleaning solvents, its melting point of 265°C was rather high, making it a difficult material to process. Moreover, its fibers often broke shortly after extrusion from the needle-fine holes of a mechanical spinnerette. "Fiber 66," as the new superpolymer was first called, needed work. But Bolton theorized that its amine and acid components might someday be cheaply obtained from the benzene in crude oil. As it turned out, DuPont chemists and engineers found an efficient way to make large quantities of rare hexamethylene diamine from readily available adipic acid by using a high-temperature, high-pressure catalytic process the Ammonia Department had developed at its Belle, West Virginia, plant.[19]

By 1936 DuPont was on an upswing. The company's profits, employment figures and stock value all rose steadily beginning from the time Roosevelt took office. Farsighted management, good labor relations, diversified product lines and creative research activities enabled the company to weather the worst of the Depression, while the New Deal, for all its stumbling and controversy, helped rebuild the nation's confidence. Indeed, DuPont's only reversal of fortunes came in 1938, after Roosevelt finally did what Lammot and many other business leaders had been urging him to do — reduce federal spending. The national economy promptly lurched downward again in what came to be known as the "Roosevelt recession." DuPont's sales volume fell 17 percent, net earnings dropped 40 percent, and the dividend on stock was cut nearly in half. Real recovery came only after the nation had begun preparing for the Second World War.

Despite its general prosperity during the Depression — indeed, partly because of it — DuPont was stung badly in the public arena in 1934 and 1935. A number of circumstances combined to bring unwelcome publicity to the company and to several of its leaders. Among these were the rise of militaristic regimes in Europe and Japan, which raised alarm among many Americans about being drawn into another war. Public attention once again focused on DuPont, which had profited from munitions sales in the First World War, even though munitions had long since shrunk to a very small portion of the company's diversified product base. Also, John Raskob, Alfred P. Sloan Jr. and Lammot, Irénée and Pierre du Pont played highly public roles in funding the conservative Liberty League, which fought bitterly to unseat Roosevelt and undo the New Deal. In the overall context of the Great Depression, these factors made DuPont appear to be an insensitive corporation that grew fat as others starved.

DuPont leadership was summoned in 1934 to defend the company against charges of wartime profiteering before the U.S. Senate Munitions Investigating committee chaired by Senator Gerald Nye.

Big business in general was under heavy attack from progressives like physician Francis Townsend in California and populists like Senator Huey Long in Louisiana. DuPont was singled out for special criticism by Detroit radio demagogue Father Charles Coughlin, who played on his listeners' fears and prejudices to enlarge his following. But hostile scrutiny also came from quarters harder to dismiss. In late 1934 Pierre, Irénée and Lammot were summoned before the United States Senate Munitions Investigating Committee, chaired by the North Dakota isolationist Republican Senator Gerald Nye, to answer charges of profiteering during World War I. Before a defense could be mounted, DuPont was tagged with the reputation of being "merchants of death."

DuPont had suffered similar aspersions during the war but had emerged with near universal praise for its efficient and expert operations. Things were different in the 1930s. Many Americans struggling to make a living and fearful of another European war were ready, even eager, to believe charges directed against DuPont and other corporations. As the Nye Committee hearings dragged on through 1935, DuPont faced a public relations fiasco that it deeply resented and was unprepared to meet. The first situation aggravated the

CHEMISTRY
BETTER THINGS FOR BETTER LIVING THROUGH CHEMISTRY

second. Lammot especially resisted giving in to the irrationality and emotion of the public relations crisis. Walter Carpenter Jr., however, quickly grasped the problem and the need for action. "Why in hell don't we do something?" he exclaimed in frustration to a fellow member of the Executive Committee in February 1935. "Nothing we can do will be 100 percent right. Doing nothing is 100 percent wrong."[20]

Between the First World War and the Nye Committee hearings, DuPont's product mix had changed from 97 percent explosives to 95 percent non-explosives. Yet the company's image in the public eye remained that of a munitions maker — an image wholly out of line with reality. Bruce Barton, of the New York advertising firm Batten, Barton, Durstine & Osborn, Inc., approached Lammot in May 1935 emphasizing that "the better your public standing as a peace-time manufacturer, the quicker the public acceptance of every new product."[21] Barton proposed a $500,000 campaign to change DuPont's image.

With mixed feelings, Lammot and the Executive and Finance Committees told Barton to go ahead. Lammot preferred the sort of public relations that came with the company's exhibits on the boardwalk at Atlantic City, and at trade shows and expositions such as the 1935 National Cotton Show in Memphis. Engineer Chaplin Tyler, then working in DuPont's Publicity Department, helped counter Lammot's opposition to broad-based public relations by urging an informative program of "technical publicity" aimed more at peoples' heads than their hearts.[22] Barton's sophisticated campaign simply bypassed both of these approaches. Barton recognized, as did politicians like Senator Nye, that people's hearts have a good deal to do with what they think, and that "facts" have a good deal to do with how they feel.

One result of Barton's public relations concept was DuPont's famous motto, "Better Things for Better Living ... Through Chemistry," which originated in BBD&O's offices and appeared in a fall advertising campaign in the *Saturday*

OPPOSITE: **This mural was painted for the 1939 New York World's Fair. It illustrated the new motto, "Better Things for Better Living ... Through Chemistry." The painting is an allegorical representation of the benefits that industrial applications of chemistry have brought to humanity.**

The mural, by Brandywine School artist John W. McCoy II, was moved in 1940 from the World's Fair to the Nemours Building in Wilmington. In 1999 it was donated to Hagley Museum and Library.

TOP and MIDDLE: **Over 26 million people passed through DuPont's Atlantic City boardwalk exhibit, open from 1916 to 1955. In 1948, the exhibit proclaimed "Chemical Research Contributes to Automotive Progress."**

BOTTOM: **DuPont displays at Atlantic City informed visitors about the workings of the large company, including this chart of DuPont's 10 industrial departments.**

NOTABLE

> DuPont decided not to make "nylon" a trademark but to keep it as a generic product name. Just two years earlier the company had lost a lawsuit against the Sylvania company, which referred to its own moisture-proof wrap as "cellophane." DuPont was unable to persuade the court that it had taken sufficient pains to protect "cellophane" as a trademark; instead, it had allowed the name to pass into common usage, as King-Seeley had done with "thermos" and Bayer with "aspirin." Indeed, the term "nylons" so quickly became the popular way to refer to women's hosiery that retaining the name as a trademark would have meant enormous effort and expense for DuPont, with no assurance of final success in the courts.

> The public had to wait an entire year after Stine's October 27, 1938, announcement of nylon hosiery at the New York World's Fair to try on the new stockings. The first sales were made in Wilmington department stores. Full-page ads in the Wilmington, Delaware, *Journal-Every Evening* invited the "Women of Wilmington" to "inspect and purchase the stockings everyone has been talking about." Needing proof of a local address in order to buy the three pairs allowed per person, women traveled from all over the country to secure a hotel room inside the city limits. The entire run from the Experimental Station's pilot plant flew off the shelves. The response, closely monitored by DuPont, was positive and led to the first national sales, called "N-Day," on May 15, 1940, in selected major markets. Even with a quota of one pair per person, the entire stock of 5 million stockings was gone at the end of the day.

> DuPont expected great things from its new Fiber 66. Eager to give the product a catchy name, the company appointed a special committee to screen suggestions. The Rayon Department's Dr. Ernest Gladding must have had tongue in cheek when he offered "Duparooh," for "DuPont Pulls A Rabbit Out Of the Hat." Other ideas were "Wacara," a tribute to Wallace Carothers; "Delawear," Lammot du Pont's favorite; and 350 other creations, like Dusilk, Moursheen, Rayamide and Silkex. After Gladding's second submission, "norun," was rejected because the new fabric *did* run, the naming committee, composed of Gladding, general manager Leonard Yerkes, and his assistant, Benjamin May, settled on the prefix "nu." The second syllable, however, remained a problem. "Nuron," a flip-flop of "norun," sounded too much like neuroanatomy. The determined Gladding then struck out the "u" and the "r" and substituted an "i" and an "l." But "nilon" could sound like "neelon" or "nillon," so "y" went in for "i" and "nylon" emerged as the winning name of what DuPont was sure would be a prodigy in its product line.

> DuPont's Chemical Department head, Elmer K. Bolton, noticed in 1930 that some companies were hiring university researchers as consultants and then covering their work with patents if there were any possibility of commercialization. "There has never been a time when competition is so keen in research as it is today," he said. DuPont also hired research consultants such as the University of Illinois's Roger Adams and Carl Marvel, who consulted at DuPont laboratories for 40 and 60 years, respectively, and steered many promising new chemists toward careers at DuPont.

CHEMICAL HERITAGE FOUNDATION IMAGE ARCHIVES

> DuPont dropped its DuPrene trademark in 1936 and referred to its synthetic rubber afterwards as "neoprene." DuPont made only the material itself and not the many products, such as insulated electrical wire, hoses and shoe soles, made by the manufacturers who purchased DuPrene and then shaped it for their own uses. DuPont marketers such as Ernest Bridgwater feared that the company would not be able to control the quality of the actual end-product that reached consumers. Under those circumstances, the generic term "neoprene" was more appropriate to DuPont's role. A more colorful reason for withdrawing the trademark, according to the Organic Chemical Department's Oliver M. Hayden, was the complaint by a West Coast entertainer that DuPont was infringing on her stage name, Duprene. "Perhaps she thought DuPont would buy her off," Hayden speculated. By that time, however, DuPont had already withdrawn the trademark.

> To save jobs during the Depression, DuPont spread out work and reduced the working hours of its wage earning employees. The company also adopted a five-day week for its salaried employees and cut their compensation 10 percent, while maintaining the regular rates paid its hourly workers. In 1934 the company granted all employees with more than one year's service an annual, paid, one-week vacation, one of the most progressive plans in American industry.

> In 1925 Notre Dame chemistry professor Father Julius A. Nieuwland announced that he had discovered a catalyst that joined acetylene into polymer chains, or macromolecules. DuPont chemists recognized a connection with their own research on synthetic rubber and visited Father Nieuwland at his lab in South Bend, Indiana. Three years later DuPont arranged for Nieuwland to serve as a research consultant. When Arnold Collins discovered DuPont's synthetic rubber, later named DuPrene, in 1930, the company acknowledged Father Nieuwland's contribution by offering to handle his patent applications in return for an option on them. Father Nieuwland had taken a vow of poverty and could not accept a consulting fee. Instead, DuPont gave $1,000 worth of journal subscriptions annually to Notre Dame.

Evening Post, and on the October 9, 1935, debut of a DuPont-sponsored weekly radio show titled "The Cavalcade of America." Featuring upbeat, educational presentations about American life and history, "Cavalcade" avoided political controversy, mentioned DuPont products in a low-key but positive way, and eventually overcame the doubts that Lammot and others had initially held. The show moved to television in 1952 and ran until June 4, 1957. It was widely praised for its quality presentations, but in 1956 when Bruce Barton summed up the program's positive effects on DuPont's image, he thought "Cavalcade" would have to share credit with another major contributor — nylon stockings, whose chemical ancestors at DuPont went back even further than Barton's public relations campaign.

On April 30, 1936, Wallace Carothers was elected to the prestigious National Academy of Sciences, the first scientist from an industrial research department ever to receive the honor. A month later he was hospitalized in Philadelphia for treatment of severe depression. Bolton transferred the supervision of Carothers's lab to George Graves, but the development of Fiber 66 had been taken up by dozens of chemists and engineers under Bolton, Yerkes and the Ammonia Department's Roger Williams. Over the next four years the project drew in more than 200 technical professionals. Crawford Greenewalt, a young chemical engineer who had joined DuPont in 1922 after graduating from MIT at just 20 years of age, was now put in charge of the engineering end of Fiber 66. He immediately felt Executive Committee pressure to "get on with it." Colleagues discovered him late one night dressed in a tuxedo in the Fiber 66 pilot plant at the Experimental Station; he had stopped by after a black tie affair to check up on things. The Committee's eagerness for progress was understandable, but Greenewalt and his coworkers still had major engineering problems to solve.

First, the superpolyamide had to be made with absolutely pure ingredients and melted at a precisely

OPPOSITE TOP: **The "Cavalcade of America" radio shows featured a full chorus made up of DuPont workers.**

OPPOSITE MIDDLE: **DuPont hired a historian, Dr. Frank Monaghan (left), a novelist, Carl Carmer (center) and a biographer, Marquis James, to research and write material for the 30-minute, weekly "Cavalcade" radio dramas.**

OPPOSITE BOTTOM: **"Cavalcade" honored doctors, military heroes, inventors and other notables. Fanny Farmer, "kitchen scientist," was the subject of a 1947 episode with music director Robert Armbruster and actress Ida Lupino.**

DuPont used the "Cavalcade of America" shows to advertise its non-explosive products as well as to celebrate ingenuity and patriotism.

TOP: **Neoprene, the only non-metal on the government's World War II "most critical" list, was first made in Louisville, Kentucky.**

LEFT: **Cordura® rayon yarn was used for tire belting. Here, at the Spruance plant outside Richmond, Virginia, strands are being drawn from a creel and wound up on a beam. In a textile mill the yarn will be twisted into cord and then shipped to a tire manufacturer.**

RIGHT: **The toothbrush had the honor of being the first item to benefit from the invention of nylon. Before its invention, the world relied on wild swine from Siberia, Poland and China for toothbrush bristles. In the 1950s, DuPont registered its nylon bristle worldwide as Tynex®.**

controlled temperature to avoid decomposition. It was then filtered, pumped into a spinnerette, spun out into threads, "cold-drawn" and wound up evenly, "sized" or covered with a material to protect it during the various mechanical operations of textile production, and, finally, dyed. Every stage posed problems, such as bubbles that formed in the pump and caused decomposition of the fibers in the spinning process. Coffman and Graves solved this problem by pressurizing the pump, which kept the gases dissolved in the molten fiber. Efforts to spin Fiber 66 taught the engineers that it was, as technician Joseph Labovsky put it, "a fickle fiber." After three years of trying, Labovsky and others were able to spin it for a grand total of 10 minutes.[23] But by May 1937 continuous spinning times had increased to 82 hours.

The triumph of Fiber 66 coincided with the tragedy of Wallace Carothers, who checked into a Philadelphia hotel room early in the morning of April 29, 1937, and, some time in the next 12 hours, swallowed the lethal dose of cyanide he had carried for years. Bolton later said that "Carothers read from the depth of organic chemistry such as I have never seen."[24] What the renowned chemist read from his own personal depths remains a mystery, though many of his friends thought he feared above all that he would never have another good idea. Carothers poisoned himself just two days after his 41st birthday, as much a victim of chemistry as its master.

Though Carothers's greatest commercial achievement, his "synthetic silk," was yet to be realized, the Wilmington papers noted his success with synthetic rubber, which DuPont continued to improve after its initial commercialization in 1932. DuPrene's noxious odor was eliminated through the use of a water emulsion process licensed from the German dyestuffs industry consortium I. G. Farben in 1934. That year Ernest Bridgwater and Vic Cosler of Orchem launched a creative promotional sales campaign aimed at developing new markets for DuPrene. Their program was so successful that by 1939 DuPrene accounted for 78 percent of U.S. domestic synthetic rubber consumption. By 1941, war-related needs required the opening of a new plant in Louisville, Kentucky, that boosted DuPont's annual neoprene production to 10,000 tons.

As Senator Gerald Nye railed against the "DuPontocracy," the company introduced a new, tough rayon product, Cordura®, for automobile tire belts; started construction of a titanium dioxide pigment plant at Edge Moor, Delaware; and broke ground at the Experimental Station for a completely new venture in its Medical Division, the Haskell Laboratory of Industrial Toxicology. Both facilities opened in 1935, though in 1953 Haskell Lab moved to new quarters in Newark, Delaware.

Other developments underscored DuPont's ongoing research success and business vigor despite the Depression and the Nye Committee hearings. Workers' wages went up 17 percent in 1936, and for the first time workers enjoyed two-week paid vacations. Though some dismissed them as "company unions," DuPont's "Works Councils," established in 1933, peacefully traversed the troubled labor-management terrain of the Depression years and contributed to DuPont's overall success. The company's new Plastics Department introduced the clear acrylic polymer Lucite® in 1936 to compete with Rohm and Haas's Plexiglas®.

DuPont's new, fully air-conditioned Nemours

"Have you got socks made out of coal, or was somebody ribbin me?"

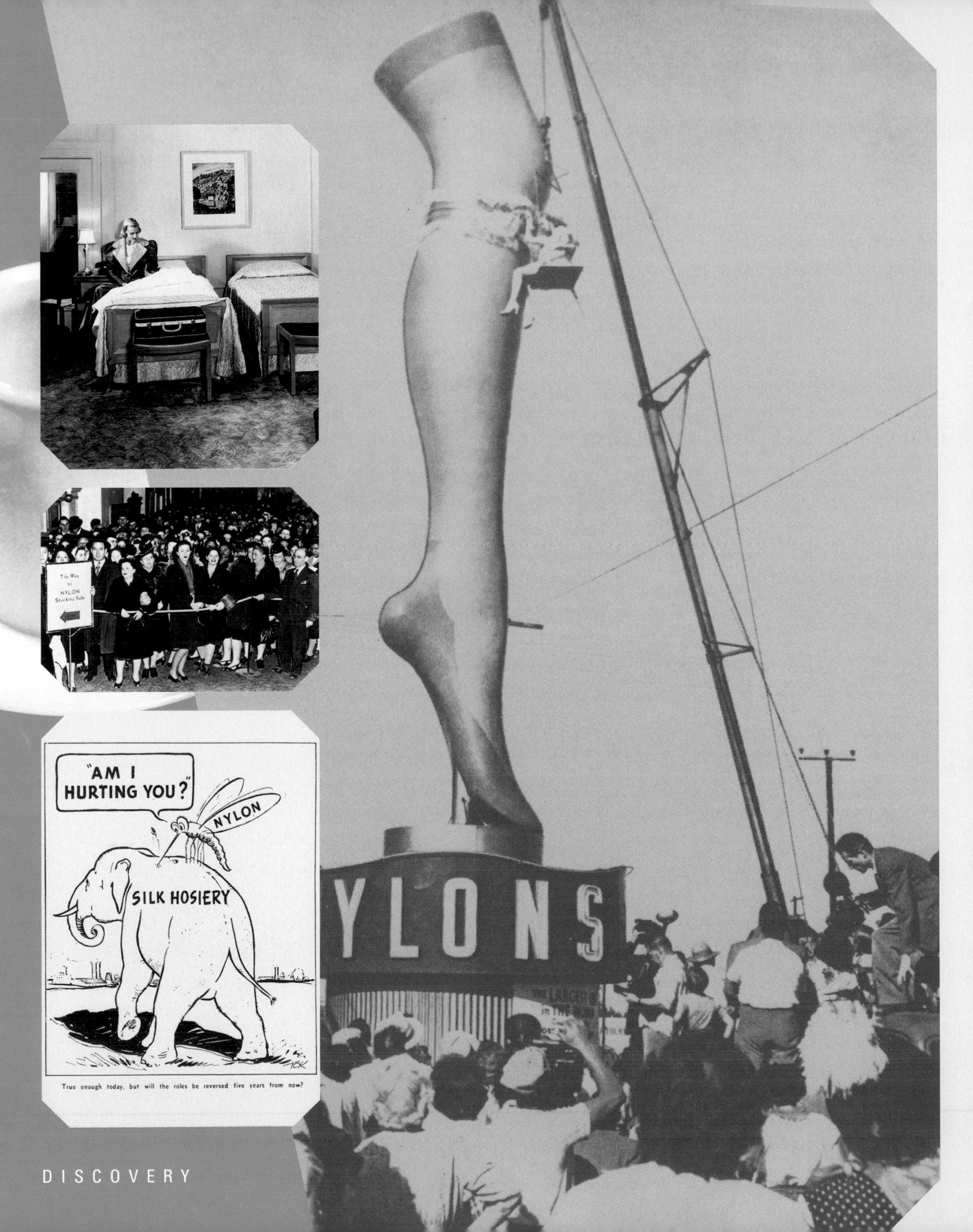

OPPOSITE LEFT: Three nylon production steps: drawing and twisting, spinning in a pressurized chamber, and sizing.

OPPOSITE RIGHT TOP: DuPont volunteers tested the effects of nylon on the skin by wearing patches.

OPPOSITE RIGHT MIDDLE: Lucky shoppers at Braunstein's Department Store in Wilmington lined up to buy nylon stockings in 1939.

OPPOSITE RIGHT BOTTOM: "Have you got socks made out of coal, or was somebody ribbin me?"

TOP: The Hotel du Pont in Wilmington helped promote nylon with this "Nylon Suite" containing nylon drapes and nylon fabrics for furniture.

MIDDLE: Shoppers on May 15, 1940, finally got their hands on nylon stockings across the United States.

BOTTOM: "Am I hurting you?" says nylon fly to silk hosiery elephant.

Actress Marie Wilson, perched atop a crane, watched as a 2-ton, 35-foot high cast of her leg was unveiled in Los Angeles to promote the sale of nylons.

Building in downtown Wilmington was completed in 1938 to house an expanded central headquarters staff that now numbered 3,200 employees. It was connected by an eighth floor walkway to the original DuPont Building, which had been expanded between 1907 and 1934 to encompass an entire city block. Butacite®, a new polyvinyl butyral plastic, was introduced and was immediately adopted by auto makers as the preferred interlayer in windshield safety glass because of its ability to withstand freezing temperatures. But DuPont's warmest enthusiasm for a new product was shown in July 1938 when the pilot plant at the Experimental Station produced the first run of Fiber 66, now named "nylon."

After allocating $8.5 million for a new nylon plant at Seaford, Delaware, and $2.5 million for a nylon intermediates (hexamethylamine diamine and adipic acid) plant at Belle, West Virginia, the company dispatched Charles Stine to the 1939 New York World's Fair to make the first public announcement of DuPont's latest offering. On October 27, 1938, Stine appeared before a meeting of women's clubs at the Fair and declared nylon to be "strong as steel, as fine as a spider's web, yet more elastic than any of the common natural fibers."[25] Silk hosiery tended to lose its shape and was notoriously subject to runs. "Nylons," as the sheer hosiery came to be known generically, were not indestructible, but when they finally went on sale to the general public in May 1940 they earned DuPont $3 million in profit in only seven months — enough to pay off the Rayon Department's entire R&D bill for Fiber 66.

Nylon was a remarkable product developed at a fortuitous time. The Seaford plant came on line on December 12, 1939, as World War II reduced, then cut off entirely, supplies of Japanese silk. Moreover, DuPont's Rayon Department was experienced in the problems of textile production, and its Ammonia Department was skilled in the complex processes required to make nylon's ingredients. These DuPont advantages would become national assets as

JOURNAL-EVERY EVENING

State Edition

Evening Founded 1871 / Journal Founded 1888 — Evening Journal and Every Evening Consolidated Jan. 1, 1933.

Wilmington, Delaware, Wednesday, October 19, 1938 — 24 Pages — Price Three Cen[ts]

Seaford Selected as Site For Huge $7,000,000 DuPont Co. Textile Mil[l]

Plant Will Be Erected on 340-Acre Plot Ab[out] Mile Southwest of Sussex County Town; To[wn] On Northern Side of Nanticoke River Adjac[ent] To the Cambridge Branch of Pennsylvania R[ail]road; Construction Will Give Jobs to Hundr[eds]

Building of Big Factory Is Expec[ted] To Begin Promptly, Announcement S[ays]

Governor McMullen Pleased Delaware Has B[een] Selected for DuPont Company's Project; Sees I[t] Tremendous Impetus in Industrial Growth of St[ate]; Dover Mayor Says It Will Prove to Be Very Benef[icial]

Seaford, Del., has been selected as the site for the e[rec]tion of a $7,000,000 textile yarn plant, officials of the duPont de Nemours & Company announced today.

The plant will be erected on a 340-acre site about a [mile] southwest of Seaford on the northern side of the Nant[icoke] River, adjacent to the Cambridge branch of the Delaware [Divi]sion of the Pennsylvania Railroad.

The text of the announcement by the DuPont Com[pany] follows:

"The E. I. duPont de Nemours & Company today [an]nounced that it had definitely chosen a site in lower Dela[ware] for the erection of the new textile yarn plant recently [au]thorized.

"The site is about a mile southwest of Seaford, De[l., on] the northern side of the Nanticoke Rver, adjacent to the [Cam]bridge branch of the Delaware Division of the Pennsyl[vania] Railroad. It embraces 340 acres."

Project To Employ Hundreds

Work on the construction of the plant will give employment to several hundred workers for about a year. The plant, when put in oper-

New Plant to Benefit This Area

DuPont Company Decision

Orders Indic

Lindy Gets Germany's Second Highest Honor

American Flier Decorated for Services to Aviation In Name of Hitler; Corresponds to France's Legion of Honor Award

BERLIN, Oct. 19 (INS).—Col. Charles A. Lindbergh, famous American airman, has received the second highest honor which the German government can bestow, it was revealed today.

At a dinner at the home of Ambassador Hugh R. Wilson last night, it was learned, Field Marshal Hermann Goering, in the name of Chancellor Hitler, awarded to Lindbergh the order of the German eagle, with a star. The order was especially created by the Nazi regime for distinguished foreigners.

This honor in Germany corresponds to that of being named a grand officer of the Legion of Honor by France.

Recognizes Aviation Services

It was explained that the decoration was bestowed upon Lindbergh in recognition of his services to aviation.

Lindbergh, who recently was the center of an international wrangle over his alleged activities during the European crisis, is the only American who has received this order. Also he is the only American who has received any decoration at Hitler's direct command.

The controversy in which the airman's name was involved arose from charges that he informed the British government at the time of the European war scare that the German air force could defeat the British, French and Russian forces combined.

The order received by Lindbergh is worn on the left side of the chest with a ribbon around the neck.

Shown How To Wear It

When the colonel received the order, he flushed, smiled and seemed obviously flustered as if he was unaware, for the moment, of exactly what had happened. Then he shook hands with Goering.

Goering then carefully explained the decoration to Lindbergh and showed him how it should be worn. Lindbergh, who arrived in Germany last week, inspected a number of aviation factories outside Berlin.

Many Letters Are Returned In Vote 'Purge'

[L]ynch Says Return Slips Will Be Scanned; Won't Give Evidence to G. O. P.

Large numbers of registered letters [se]nt out by the Democrats to "purge" [al]leged phantom voters from the [re]gistration lists have been returned "undeliverable," Democratic officials [sa]id today.

Stewart Lynch, Democratic candi[da]te for attorney-general, who had [no]t yet been at his office, said today [he] was informed the first batch of [ap]proximately 200 letters which had [be]en mailed resulted in 150 being [ret]urned as "undeliverable" while 35 [ret]urn receipts were received.

Mr. Lynch, from whose office in [th]e Equitable Building the registered [let]ters are being mailed, stated that [th]e 35 return receipts will be checked [car]efully and if they are found to [be] satisfactory no further steps will [be] taken in those cases. "If they [are] not, however," Mr. Lynch con[tinu]ed, "they will be turned over to [the] Postoffice Department and to [the] office of the district attorney."

Sent to Front Street

[T]he first batch of letters sent out, [Mr.] Lynch said, was addressed to [vote]rs registered from the Pennsyl[van]ia right-of-way on Front Street, [vari]ous warehouses and buildings [torn] down on Shipley, Orange and [...]all Streets.

U. S. Deserter Shown Up As An Inept Spy

Failure as Dishwasher, He Had Bungling Career As Nazi Secret Agent, He Testifies in Court

NEW YORK, Oct. 19 (AP).—Guenther Gustav Rumrich, an unsuccessful dishwasher who became a $40 a week "mail order" spy, continued in Federal Court today his story of a bungling career as agent for a German military espionage ring.

Rumrich, 32-year-old U. S. Army deserter, related yesterday he had brazenly sold matters of public knowledge to German officers who believed they were buying vital defense secrets.

Failed to Produce

Rumrich—a modern Benedict Arnold who failed to produce—from the witness stand identified Johanna Hofmann, 26, one of three codefendants, as a messenger for the espionage syndicate.

Fraulein Hofmann, buxom former hairdresser on the German liner Europa, sat calmly chewing gum as Rumrich testified he had turned over to her photographs of the U. S. Navy cruiser Houston in the Panama Canal locks for delivery to the syndicate's Berlin headquarters.

Courtroom spectators laughed as the government's star witness, who said he wrote to a German newspaper to get his job as spy, admitted he had bought the "secret" photographs in a store in the Panama Canal Zone.

Vague About Plot

Rumrich also told of a plot to lure Col. Henry W. T. Eglin, commandant of Fort Totten, N. Y., to a Manhattan hotel and gain from him the U. S. Army's mobilization plan for coast artillery on the Atlantic seaboard.

Asked how he and Kiri Schlueter, a fugitive defendant identified by U. S. Attorney Lamar Hardy as the ring's "contact man," planned to overcome Colonel Eglin, the witness said vaguely:

"By some kind of a fountain pen with some kind of a gas that would

Police Gun Ends Life of Monkey

OPPOSITE TOP: **Nylon was first produced at the Seaford, Delaware, plant in 1939.**

OPPOSITE BOTTOM: **It was front page news when Seaford, Delaware, was named as the site of the first nylon plant. The town calls itself "Nylon Capital of the World."**

LEFT TOP: **In addition to windshields, Butacite® found application in safety windows, glass doors, bathtub enclosures, shop windows, skylights and tabletops.**

LEFT BOTTOM: **DuPont's "Lady in a Test Tube" at the 1939 Golden Gate Expo in San Francisco showed that clothing, hats, purses and stockings could all be made in the chemical lab.**

RIGHT TOP: **At the 1939 New York World's Fair, DuPont's 100-foot Tower of Research was an abstract representation of chemical research. The tower featured rising bubbles to dramatize molecular activity.**

RIGHT BOTTOM: **Opening the DuPont exhibit at the World's Fair were (l to r) Irénée du Pont, Henry B. du Pont, R.R.M Carpenter, William du Pont and Lammot du Pont.**

TOP: During World War II, nylon was used for war-related needs rather than women's hosiery. One of the first uses was to replace silk in making parachutes.

CENTER: For the DuPont World's Fair exhibit, a craftsman spent 8 days cutting an 8-pound block of Lucite® to form a perfect diamond over 9 inches in diameter.

BOTTOM: DuPont's "Test-Tube Lady" was intended to show that there was nothing to fear from science, an important message in the 1930s.

OPPOSITE TOP LEFT & OVAL: The public first saw nylon hose on DuPont models at the 1939 World's Fair.

OPPOSITE CENTER: DuPont president Lammot du Pont (right) and exhibit designer Walter Teague.

OPPOSITE TOP RIGHT & BOTTOM RIGHT: At the World's Fair, DuPont put science on exhibit to inspire visitors with the "Wonder World of Chemistry" and to assure them of the company's intent to use science to improve their lives.

war production began not far in the future. But in the meantime, women's stockings, the only commercial market large enough to justify building an entire plant for nylon production, just happened to be the ideal medicine for DuPont's "merchants of death" public relations ills.[26] What could be further removed from shipping munitions to male belligerents than offering delicate comfort and adornment to women?

The New York World's Fair officially opened on April 30, 1939, on Long Island. Among the exhibits was DuPont's "Wonder World of Chemistry," which highlighted the spinning of nylon; a $9\frac{1}{2}$-inch Lucite® diamond; a giant steel ball that crashed harmlessly into a sheet of Butacite®; Princess Plastics, a model dressed from heel to hat in synthetics produced by DuPont; and a movie titled after the exhibit's name.

The World's Fair was only four months old when Adolph Hitler's tanks overran Poland on September 1, 1939. Two days later Britain and France declared war. Publicly the United States asserted neutral status while privately Roosevelt devised ways to support the Allies without unduly alarming Americans anxious to avoid bloodshed. In the mixed atmosphere of nylon's triumph and war's approach, Walter S. Carpenter Jr. succeeded Lammot du Pont in 1940 as president of the company, the first executive not in the du Pont family to hold that office.

When the United States was finally drawn into World War II by the bombing of Pearl Harbor on December 7, 1941, the wonders of chemistry that had begun to so enhance the lives of American consumers were quickly employed to serve the necessities of warfare. Nylon stockings disappeared, replaced by nylon parachutes, ropes and tents. Before long America and the world would learn from Hitler's dark regime what it really meant to be a "merchant of death." But DuPont would also face new risks as it applied its sophisticated discovery and development skills to the ancient problem of human conflict.

WORLD OF CHEMISTRY
THIS TOWER, A SYMBOL OF MODERN CHEMISTRY, IS A GIANT REPRODUCTION OF AN EXTRACTION UNIT, AN INDISPENSABLE TOOL OF CHEMICAL RESEARCH...

CHAPTER 6

SCIENCE AND THE AFFLUENT SOCIETY

BACKGROUND: **A *Chicago Tribune* artist sketched scientists watching the world's first nuclear reaction at Stagg Field, Chicago, in 1942.**

RIGHT: **Fuel from the first nuclear reaction embedded in a plastic replica of the Stagg Field "pile."**

BELOW: **As a DuPont engineer, Crawford Greenewalt served as liaison between the scientists of the Manhattan Project and DuPont's Wilmington operations, including the Engineering Department. He became company president in 1948.**

RIGHT: **Physicist and Nobel laureate Enrico Fermi led the scientific team that constructed the first chain reaction.**

Crawford Greenewalt took his place with several other guests on a balcony overlooking the squash courts. Hollowed out under the west stands of the University of Chicago's Stagg Field, the unheated courts had been commandeered several months earlier to prepare for the day's event. On the floor below Greenewalt a 400-ton graphite block hunkered under a protective balloon of gray-smudged fabric, impassively awaiting the midwifery of physicist Enrico Fermi. Throughout the morning, Fermi and his assistants had slowly pulled 15-foot cadmium control rods from the cryptic hulk, freeing millions of submicroscopic neutrons to bombard the six tons of uranium embedded in its belly.

It was slow work. The technicians had already eaten lunch by the time DuPont engineer Greenewalt arrived in mid-afternoon on December 2, 1942, four hours after the removal of the first rod from Fermi's nuclear pile. Now the staccato "click-click" of the measuring instruments accelerated, blurred, then dissolved into a massed chatter of white noise, signaling the nuclear chain reaction roiling to a crescendo beneath the block's smooth surface. To the relief of Fermi's guests, a safety rod the assistants had nicknamed "Zip" clanged noisily into place, soaking up the runaway neutrons and shutting down the chain reaction. Greenewalt walked with his host, physicist Arthur Compton, across the University of Chicago's chilly campus. They talked excitedly not of bombs and war but of electrical power and future possibilities. Compton warmed to the glow on his friend's face. Greenewalt had seen, and appreciated, exactly what Compton had hoped he would see — not just a scientific "first" but the birth of a new age.

It was only five days to the first anniversary of Pearl Harbor, and World War II was already three years old, but DuPont had been preparing since the German invasion of Poland in September 1939. Mindful of the criticism it had received for supplying munitions to the Allies in World War I, the company, like the nation, was reluctant to take part in the latest European calamity. When France fell in June 1940, however, and beleaguered Britain stood defiantly alone beneath the blitz of German air raids, the tide of American opinion edged toward active support for the island kingdom. During 1940 and 1941 DuPont began making munitions for both Britain and the United States, now tentatively preparing for war. By the time of Pearl Harbor, DuPont workers in 14 states were helping supply TNT and other explosives for the Allies. The Seaford plant was turning out nylon for parachutes and ropes, and the company had built a neoprene plant for the government in Louisville, Kentucky. DuPont also vindicated its World War I dyestuffs research investment by offering, this time around, an abundant array of dyes for military use.

But the threat of an approaching atomic cloud overshadowed all this activity. Scientists like Leo Szilard and Eugene Wigner, who had escaped Hitler's rampaging Third Reich, warned of German progress in atomic research. After all, it was German researchers who had first split the atom late in 1938. When intelligence sources confirmed the scientists' alarm, the U.S. government quickly started its own nuclear research effort and placed it under military control.

INSET: **Dignitaries at the Hanford Engineer Works included (left to right): Vannevar Bush, director of the Office of Scientific Research and Development; James B. Conan, scientist; Brig. Gen. Leslie Groves, director of the Manhattan Project; and Franklin Matthias, military construction manager of Hanford.**

The cyclotron at the University of California's Lawrence Radiation laboratory in Berkeley helped physicists determine the properties of plutonium.

In November 1942, Brigadier General Leslie Groves, director of the top-secret Manhattan Project, sat before a skeptical DuPont Executive Committee in Wilmington. A member of the Army Corps of Engineers, he had worked with DuPont on military projects before. In his mind, it was the only company in America for the job he had taken on — producing plutonium and uranium for atom bombs.[1] But to company president and committee chair Walter S. Carpenter Jr., Groves's request raised once more the awful public relations specter of "merchants of death." It seemed preposterously risky. To vice presidents Willis Harrington and Charles Stine, it seemed not just risky but nearly impossible.

A team of DuPont scientists and engineers had just returned from Fermi's code-named Metallurgical Laboratory in Chicago, where they had gone at Groves's invitation. Charles Stine, Roger Williams, Thomas Gary and Thomas Chilton had been impressed with the Met Lab's scientific talent, including Nobel Prize winner Fermi, Leo Szilard, Eugene Wigner, Norman Hilberry and Arthur Compton, also a Nobel Prize winner, who oversaw all Manhattan Project research. However, the Brandywine veterans of neoprene and nylon found the physicists' attitude toward the problems of large-scale engineering alarmingly casual, sometimes even dismissive. Wigner, who had some engineering background but whose experience with Nazism in Europe led him to suspect all business-government dealings as subversive of scientific integrity, once fumed, "Give me some hacksaws and a couple of hammers and we'll do better than any damned DuPont Company."[2] Such sentiments did little to soften the Executive Committee response to Groves's entreaties.

Nevertheless, Groves pitched the idea hard to the Executive Committee. The consequences of German victory were unthinkable. America's success would mean a shortened war and the saving of Allied lives, not to mention the preservation of freedom. Only DuPont had the experience in large-scale military production, facility design and construction as well as the chemical engineering expertise to steer the Met Lab's wizardry to practical success. No less important, DuPont had earned Groves's trust. He felt sure that its straight-dealing engineers would not cut corners, shave the odds or cook the books.

As Groves knew it would, the Executive Committee looked past his patriotic appeals to the nuts and bolts of his proposal. Physicist Norman Hilberry accompanied Groves to Wilmington and watched the Executive Committee grill the general. "You understood why DuPont was the company it was," he said later. "They were the hardest-nosed, sharpest bunch of characters you ever ran into."[3] DuPont's strength was chemistry and chemical engineering; the company had no experience with the new element, plutonium.

Indeed, outside the few highly specialized scientific laboratories, knowledge about plutonium hardly existed. Glenn Seaborg and his team of physicists at the University of California had identified traces of a new element in less than a millionth of a gram of neptunium in the Berkeley cyclotron in January 1941. A year later Seaborg named the substance "plutonium." It remained so elusive that it could be detected only by ultra-sophisticated laboratory equipment. Now recruited to the Manhattan Project, Seaborg would try to find a way to make pounds of it. DuPont's job would be to translate his ideas into steel and concrete for full-scale plutonium production, and then to build a plant for making the uranium isotope U-235. In the race to build the bomb, both of these highly fissionable materials would be tried.

Charles Stine, usually an optimist, flatly told Groves that there was only a 50-50 chance of success.[4] In fact, the odds against achieving success before the war ended — sometime in 1945, Stine figured — were much greater. But for all its skepticism, in the end the Executive Committee could not turn Groves away. Pending approval by the Board of Directors at its next meeting in mid-December, the committee agreed to design, construct and operate a pilot plant for making U-235 as well as a full-scale plutonium production

It took just under two years for DuPont engineers and construction crews to transform 600 square miles of sagebrush desert into a plutonium production facility. Construction manager Franklin Matthias would recount years later that, for much of the project, construction ran ahead of the engineering design. Workers moved forward with rough sketches and dimensions until blueprints arrived from the DuPont design group in Wilmington.

plant. In the meantime, a committee chaired by MIT's renowned chemical engineer Warren K. Lewis and composed of DuPont's Tom Gary and Crawford Greenewalt, along with Standard Oil's Eger Murphree, would visit the Manhattan Project's research lab in Berkeley, where Seaborg and other scientists, like Ernest O. Lawrence, were exploring ways to produce U-235. DuPont's Executive Committee was willing to help the cause but wanted to make sure to put most of its eggs in the stronger basket. Groves had yet to persuade DuPont that the basket would be made of plutonium.

On their way back from California, the Lewis committee members stopped off at the Met Lab in Chicago. Their timing was propitious, for Enrico Fermi was planning an experiment for December 2 in the empty squash court. Space was limited, and Arthur Compton was allowed to bring only one guest from the Lewis committee. His choice of Crawford Greenewalt was deliberate. At 40 years of age, Greenewalt was likely to live longer than any of the others to bear witness to the historic event. Eugene Wigner brought a bottle of Chianti he had hoarded for months after the war blocked Italian imports. Wigner, who had gone to Princeton University in 1935 after being fired from the University of Berlin because his wife was Jewish, was ready to toast Fermi, who had escaped from Mussolini's Italy with his Jewish wife and their two children in 1938. When the safety rod finally clanked into place, Wigner produced the Chianti. No one offered a toast. Instead, a quiet tilting of paper cups expressed somber satisfaction at the day's achievement, the hoped-for beginning of the end of the war.

The Met Lab success on December 2, 1942, helped persuade DuPont's Board of Directors to approve the building of a pilot plant for U-235 production in Clinton, Tennessee, near Oak Ridge, and of a plutonium works on a site the company would select with the Army Corps of Engineers. However, the board stipulated three conditions: DuPont would receive only $1 above costs for the entire project; the government would take control of any patents emerging from the work; and the government would cover DuPont's liabilities and losses. Walter Carpenter Jr. would not have his company stung by accusations of war profiteering or reckless weapons work. Over the next two years he frequently reminded Walter O. Simon, DuPont's operations manager at the plutonium plant, "You've got the whole company in your hands." Simon was too young to be much worried. Later, he thought he should have been scared stiff.[5]

DuPont disguised its Manhattan Project work with a code-name, TNX, after an obsolete trinitroxylene explosive Charles Reese had developed for the Navy in World War I. Project TNX was secreted in DuPont's Explosives Department under Roger Williams. It became, in effect, the company's 11th industrial department. Williams's job was daunting. So was Greenewalt's; he was to serve as liaison between DuPont and the independent-minded Met Lab physicists. When Williams, Greenewalt and engineer Tom Gary learned of their assignments to TNX, they headed straight for the Hotel du Pont's Brandywine Room for a drink. "Lord help us," wrote Greenewalt in his diary. "Our chances of putting it over are not much better than one in four."[6]

Twenty-one months and $350 million later, on September 27, 1944, the first of three nuclear piles — DuPont engineers had taken to calling them "reactors" — was ready for start-up. Six hundred square miles of sagebrush desert and irrigated fruit orchards next to the Columbia River in the tiny hamlet of Hanford, Washington, had become the mammoth Hanford Engineer Works. Three cube-shaped reactors, each about five stories high and 40 feet on a side, were spaced at six-mile intervals along the Columbia River, to the north. They produced irradiated uranium in the form of metal-encased slugs, glowing red and containing tiny amounts of the prized fissionable byproduct plutonium. The flashlight-sized slugs were dropped temporarily into water troughs for cooling, then transferred by rail in lead-lined containers 6 to 11 miles downriver, where the plutonium was chemically separated

TOP LEFT: **"We lined up for everything," recalled one veteran of the Hanford Works. This was the drug store.**

CIRCLE and TOP RIGHT: **Once a week, the pay wagon moved from site to site at Hanford to distribute wages to construction workers.**

BOTTOM LEFT: **Uranium bars were weighed as they arrived at Hanford.**

BOTTOM MIDDLE: **The eight mess halls at Hanford served meals to as many as 45,000 construction workers daily.**

BOTTOM RIGHT: **Many Hanford workers sent money home from the Western Union station on the plant grounds.**

from the slugs. There, three massive, 800-foot-long separator buildings rose 10 stories above the desert floor. The isolation of these behemoths might have conjured up images of ancient pyramids. The men who built them chose more familiar nicknames — "Queen Marys" and "the canyons."

Inside, technicians shielded by seven feet of concrete moved the irradiated uranium by remote control through various stages of processing along the full length of the canyons. One ton of uranium slugs yielded a mere 79 pounds of plutonium nitrate, a brown, syrupy sludge that was then trucked to the Manhattan Project's lab at Los Alamos, New Mexico, for final processing into lesser amounts of pure, weapons-grade plutonium.

As they had done before at Old Hickory, DuPont crews and subcontractors built a small city at Hanford to house, feed and entertain a wartime peak of 45,000 workers. Many of those workers succumbed to the area's "termination powder," as they called it — fine sand cast loose to the winds by heavy machinery that dug up 25 million cubic yards of earth during the two years of construction. The gravelly earth around the river was one of the "acts of God" that Roger Williams credited for DuPont's success, for it made wonderfully convenient, high-quality concrete. But workers were sometimes demoralized by the ever-present grit infusing their laundry, their hair and their skin. "The turnover hazard started on arrival," said Groves.[7]

DuPont did what it could to ease the harsh conditions at Hanford. When female clerical staff complained that the rough walkways were ruining the heels of their shoes, DuPont engineer Granville Read poured asphalt. And when trouble broke out in the saloons, DuPont crews hinged the windows so authorities could restore order handily by tossing in tear gas canisters. Hanford's dust-parched construction workers gulped 12,000 gallons of beer every week. Its nuclear reactors, where the consequences of overheating were more serious, each consumed 75,000 gallons of cold Columbia water per minute.

CARPENTERS
4-801
4-1250
4--451
4--800
4--100
4--450
3--2051
3--2600
3--1601
3--2050
PAY WAGON
LABORERS

Hanford was the largest construction project of the war. The facility took up more than 71 times as much space as Old Hickory, the largest project of the First World War. Despite DuPont's concerns about associating too closely with weapons work, Hanford cemented the company's reputation as the nation's premier firm for large-scale, technologically complex projects.

Early in February 1945 the first steel cylinders of jelly-like plutonium nitrate reached Los Alamos. Soon U-235 also arrived from Oak Ridge, Tennessee. A plutonium test bomb exploded successfully at Alamogordo, New Mexico, on July 16. There was no need to test the U-235 bomb, for its detonation mechanism was better known. In Europe, the June 6, 1944, Allied invasion of Normandy bound up its many wounds and rolled onward to the liberation of Paris in August 1944. After the Battle of the Bulge, a fierce but failed counteroffensive in Belgium that bent back the Allied line of advance, Germany at last surrendered on May 8, 1945. As it turned out, sabotage by daring Norwegian resistance fighters had scuttled the Reich's atomic bomb research operation in Norway, and a general lack of resources within Germany itself had kept the Nazi nuclear effort from realizing its dangerous potential.

Yet the war in the Pacific raged on. Japan's ultimate defeat now seemed inevitable, but only after an anticipated high toll in American lives.[8] Nearly half of the 60,000 U.S. Marines who struggled across the murderous beaches of Iwo Jima in February and March had been killed or wounded — the highest casualty rate in Marine Corps history. Taking the Japanese mainland was expected to be just as bloody. When Franklin Roosevelt died on April 12, the burden of the war's final decisions settled on Harry Truman. On August 6 the B-29 *Enola Gay* released a 4-ton U-235 bomb on Hiroshima. But only silence came from Japan's military command. Was it stunned confusion or implacable defiance? Three days later another B-29, Bock's Car, dropped a five-ton plutonium bomb nicknamed "Fat Man" through a hole in the clouds over Nagasaki. The next day Japan's emperor at last pushed a

TOP LEFT and RIGHT: **American soldiers landing in France on D-Day and on the island of Iwo Jima in 1945. The heavy toll sustained in these invasions influenced the decision to bomb the Japanese mainland to bring an end to the war.**

BACKGROUND: **In September 1944, Allied paratroopers landed in Holland, many beneath parachutes of DuPont nylon.**

OPPOSITE TOP LEFT: A red parachute of high-tenacity rayon carried supplies to airborne troops.

OPPOSITE TOP MIDDLE: A DuPont operator inspected a panel of Lucite® acrylic resin for a bomber nose cone.

OPPOSITE TOP RIGHT: Many DuPont employees served in uniform during World War II.

OPPOSITE BOTTOM LEFT: DuPont employees donated blood to the Red Cross.

OPPOSITE BOTTOM MIDDLE: Gas rationing began in May 1942. Lammot du Pont rode to work in Wilmington on his bicycle.

OPPOSITE BOTTOM RIGHT: An iron wheel used to make DuPont gunpowder in the 19th century was donated to the war effort as scrap metal.

LEFT: Women who worked for DuPont during World War II were called "WIPS" — Women in Productive Service. WIPS were critical to the operation of the Seaford, Delaware, nylon plant.

New Yorkers celebrate the end of World War II in the Pacific on August 14, 1945.

OPPOSITE TOP LEFT: As DuPont president, Walter S. Carpenter Jr. guided the company's wartime production and its return to peacetime manufacturing.

OPPOSITE RIGHT: The DuPont Wilmington staff on December 6, 1945, was busy realigning production plans to meet civilian needs.

decision to surrender past the stone wall of his military commanders. After its Holocaust, its fire-bombings, and its 50 million dead, World War II shuddered twice on its nuclear funeral pyre in the Pacific, then sank into memory.

No one was more relieved than Walter S. Carpenter Jr., who reminded stockholders in a 1946 summary of DuPont's wartime activities that less than 25 percent of the company's production in the war had been military explosives, compared with 85 percent in the First World War.[9] Carpenter did not overlook DuPont's herculean munitions work — its 4.5 billion pounds of military explosives had tripled its entire World War I output — but he hastened to emphasize comforting, life-saving products: neoprene for life rafts; sulfa drug ingredients; nylon mosquito netting and surgical sutures; Freon® aerosols and refrigerants; cellophane food wrapping; x-ray film; and plastic prosthetics. Like millions of Americans, said Carpenter, DuPont had interrupted its normal activities to perform its patriotic duty. Now it was time to get back to peace and the pursuit of prosperity.

Carpenter captured the national mood precisely.[10] Often described as the greatest redistribution of income in the United States in the 20th century, the federal government's World War II expenditure of $186 billion effectively ended the Great Depression.[11] Salaries and wages more than doubled between 1939 and 1944, more than enough to offset wartime inflation. Consumers flocked to department stores, where they spent five times as much in 1944 as in 1939 on items made scarce by wartime rationing of fabrics, metals, leather, rubber, sugar, coffee and gasoline. The complaints about broken heels at Hanford had not been made lightly, for the clerks and secretaries who made them were walking in rationed shoes. By the war's end consumers had acquired $140 billion of spending power, nearly triple the amount available when Pearl Harbor was attacked. Additionally, the tuition and mortgage benefits of the GI Bill eased the transition to civilian jobs and housing for nearly 13 million veterans. Within two years the ex-GIs and their families constituted nearly a quarter of the U.S. population.

The war had favored large businesses, despite a modest effort by the federal government to distribute a share of lucrative military contracts to small firms. The War Production Board allotted 30 percent of its contracts to the 10 largest corporations and promised immunity from antitrust prosecution until the end of the war. As the Manhattan Project showed, some tasks were simply too big and some deadlines too important to risk the inefficiency of job-splitting and the plodding pace of scrupulous equity. Still, Assistant Attorney General Thurman Arnold of the Justice Department's Antitrust Division indicted DuPont and the American Lead Company in 1942 for conspiring to restrain trade. DuPont settled quietly for a small fine without admitting guilt but in 1943 was slapped with two more charges, one involving titanium dioxide pigments and the other its long-standing patent-sharing agreement with England's Imperial Chemical Industries, Ltd. Prosecution of these cases was postponed until the war's end.

All told, Arnold filed an astonishing 180 cases — half of all the antitrust cases ever brought under the Sherman Act of 1890 — before Roosevelt neutralized him in 1943 with an appellate court appointment. However, the

BELOW: **The DuPont/General Motors antitrust case was litigated for 12 years. This was a courtroom scene in January 1955.**

OPPOSITE: **DuPont brought busloads of high school students to the Experimental Station in 1951 for "Exploring Business Day." Students were encouraged to pursue a science education by viewing scientists at work.**

end of the war triggered not only resumption of the prosecution of the pending suits but also new charges against DuPont over paints, cellophane, wood finishes, brake fluid and its General Motors stock. In 1951 DuPont faced six major antitrust lawsuits. The company won two, including one before the U.S. Supreme Court (the cellophane case), and two more were dismissed. In 1952 the patent-sharing case was decided against the company and ICI. The two companies were forced to end their agreement and divide their shared assets in Brazil, Argentina and Canada. In the GM suit, federal prosecutors charged that DuPont's 22 percent share of GM holdings gave it an unfair "inside track" to sell its many auto-related products.

The litigation and settlement of the GM case dragged on for 12 years. Though DuPont initially won the case in 1954 before the Federal District Court in Chicago, the U.S. Supreme Court ruled against DuPont three years later, not because DuPont and GM had conspired against competitors but because DuPont's GM holdings, in and of themselves, created a "reasonable probability" of preferential treatment. Over the next four years a settlement was hammered out with great difficulty, ensuring disposal of DuPont's 63 million shares with minimal tax penalty to shareholders.

DuPont was not the only company whose antitrust cases reached the Supreme Court between 1945 and 1964, but DuPont went before the Court more often than any other company. DuPont was involved in three of the Court's 20 major antitrust rulings in the postwar period, winning its cellophane case but losing twice on matters relating to General Motors.[12] Sensitive to its historical reputation as a good corporate citizen and a solid stock investment, DuPont was stung sharply by the GM case and by other antitrust actions through the 1940s. In 1947, Carpenter's indignation spilled over into the company's staid Annual Report. "It may well be," he wrote, "that the DuPont company, as well as other companies, may have to reconsider their time-honored policy of making new and better products available to the public at lower prices."[13]

Carpenter, who had once told a young DuPont engineer wavering over an unwanted assignment, "Young man, you ought to accept challenges when they're offered to you," turned the challenge of the company presidency over to that same man on January 19, 1948. Crawford Greenewalt had helped pilot DuPont's nylon R&D triumph, had won over the Met Lab's dubious physicists to the team effort central to the Manhattan Project, and then had helped solve several complex design problems to ensure that project's success. Now, as DuPont's new president, Greenewalt would oversee an unprecedented period of expanded research and production while defending the company's increasing size as helpful, not harmful, to individualism and free enterprise.[14] Federal antitrust action sounded a discordant drone beneath DuPont's otherwise seemingly synchronized harmony with the American economy in the 1950s.

DuPont's success had always been a calculated gamble, especially after the company committed itself to Charles Stine's fundamental research program in the late 1920s. By 1950 research was no longer a choice for a competitive chemical company — it was a necessity. "We do research because we have to," said Roger Williams, who succeeded Stine as research advisor to the Executive Committee in 1945. "If we let up, our competitors would trim us."[15] Greenewalt, who had asked the Lord's help when facing 1-in-4 odds in the Manhattan Project, now proceeded to spend $30 million for expansions at the Experimental Station with only a 1-in-20 chance that any project hatched there would ever pay off. During 1951, when the Experimental Station's 19 new buildings opened, DuPont spent $47 million on research, several million dollars shy of the R&D budgets of such giants as General Motors and Bell Telephone but still more than double the amount spent by Standard Oil. By 1956 each of the 10 DuPont manufacturing departments had a long-range research facility on the Experimental Station campus.

The risks of such enormous investments were clear,

but so were the calculations behind them. Diversification maximized the potential for a profitable product, as did careful review by departmental research committees, backed up by an elaborate statistical tracking of the company's return on investment. Every month the Executive Committee gathered to review 350 business performance charts swung before them on overhead trolleys. Luck and intuition also played a role, and for many years both had favored the company. After all, three-quarters of DuPont's products in 1947 had been introduced in the 20 years since Stine's basic research initiative. Now DuPont opened the cornucopia of science once more into the hands of an increasingly affluent society.

Out came Orlon®, an acrylic, wool-like fiber that DuPont concluded could be dyed properly even as the May plant in Camden, South Carolina, was being constructed during 1949. "When you look back on it," recalled Executive Committee member Lester Sinness, "we were Mississippi river boat gamblers."[16] Orlon® presented other technical problems as well, such as drawing and cutting the filament to maximize its "woolliness." But nearly $60 million later the gamble began to pay off in the bulky sweaters that had just come into style. In 1956 DuPont doubled the May plant's capacity and started building a new textile fibers plant in Waynesboro, Virginia.

DuPont's next offering was the wash-and-wear polyester Dacron®. The particular polymer that became Dacron® polyester had been prepared as early as 1934 by Edgar Spanagel, a chemist working with Wallace Carothers. But DuPont had shelved the discovery to concentrate on more promising nylon research also underway. Six years later two British researchers familiar with Carothers's publications also discovered the polymer, patented it as Terylene, and sold a 20-year license to ICI. By that time, however, DuPont had resumed its polyester research and, early in 1945, purchased the U.S. rights to Terylene from ICI. With that roadblock removed, DuPont proceeded with its ongoing development of what it called "Fiber V."

OPPOSITE TOP and BACKGROUND: **Bobbins of Orlon® yarn being inspected at the May plant, Camden, South Carolina. The plant opened in 1949 and, after Orlon® proved extremely popular, capacity was doubled in 1956.**

OPPOSITE CIRCLE: **"This classic sweater set of Orlon® acrylic fiber feels amazingly soft," declared a 1951 ad.**

OPPOSITE MIDDLE LEFT: **The Executive Committee met monthly in the Chart Room to review company performance.**

OPPOSITE BOTTOM LEFT: **Company President Crawford Greenewalt (standing center) with four former presidents in 1949 (left to right): Walter S. Carpenter Jr., Pierre S. du Pont, Irénée du Pont and Lammot du Pont.**

TOP: **Uniforms of Dacron® quickly became popular because they were durable and did not need ironing.**

CIRCLE: **This quick-drying bathing suit was made of Dacron® and cotton.**

BOTTOM: **Dacron® was made first in a pilot plant in Seaford, Delaware, then went into full production in Kinston, North Carolina.**

COLUMN ONE: **Production of nylon returned to yarn for hosiery and other peacetime products after World War II during which it had been used for parachutes, mosquito netting and surgical sutures.**

COLUMN TWO: **A San Francisco street was packed with shoppers in 1946 for the first postwar sale of nylon hosiery. DuPont took out an ad that year to let consumers know there would be a delay in converting nylon production back to civilian markets. DuPont celebrated the 20th anniversary of the discovery of nylon in October 1958 with a cake decorated with bobbins. Julian Hill blew out the candles.**

COLUMN THREE: **Polyester and nylon weren't restricted to fibers. Cronar® film was ideal for graphic arts applications, and Mylar® was strong enough to hoist a 3-ton car. Tough, lightweight Zytel® nylon resin quickly became a standard material in tools and machines.**

COLUMN FOUR: **Cronar® polyester film was made at Parlin, New Jersey. Freon® propellant powered new aerosol sprays. Delrin® and other engineering plastics replaced metal in automobile components.**

In 1950, a pilot plant at Seaford, Delaware, produced Dacron® fiber with modified nylon technology. DuPont opened its huge Kinston, North Carolina, plant in 1953 to produce Fiber V, or Dacron®. Like Orlon®, Dacron® posed problems that were not always evident until the product was tried out by potential customers. For instance, tire manufacturers found Dacron® disappointing as a tire cord, DuPont's first intended use for it. When woven as a textile, Dacron® tended to form "pills," little clumps of fiber on the surface of clothing. So in 1954, following Roger Williams's recommendation, DuPont opened its Chestnut Run Textile Research Laboratory near Wilmington to explore such "end use" problems with synthetic fibers. By 1960 the company was operating a second Dacron® plant, next to the rayon plant at Old Hickory.

Nylon, of course, came tumbling out in record amounts to meet a huge renewed demand for stockings. By 1950 five of DuPont's 10 manufacturing departments were involved in producing synthetic fibers or the chemicals required for their production. Moreover, renewed military demands during the 1950-1953 Korean War strained DuPont's nylon and rayon production capacity despite a temporary slump in the nationwide textile market in 1952. In 1949 DuPont's overall sales hit $1 billion for the first time. In 1951 they reached $1.5 billion; in 1953, $1.75 billion; and in 1957, the company posted record sales just short of $2 billion.

Diversification at DuPont meant more than just new synthetic fibers. Research yielded an array of new plastics, like Delrin® and Zytel®, strong enough to substitute for metal in gears, tools and automobile parts; agricultural products, such as fertilizers, fungicides, herbicides and seed disinfectants; and two offspring of Dacron®, the polyester films Mylar® and Cronar®. Mylar® appeared in various products such as magnetic recording tape and specialized packaging, and Cronar® made a durable, nonflammable motion picture film as well as engineering and graphic arts films.

At Textile Fibers' Pioneering Research Lab at the

New York Times February 6, 1946

Yesterday Macy's sold

50,000 pairs of nylons...

★

An apology to those

who didn't get theirs...

Yesterday, for the fourth time since early November, Macy's put nylons on sale. We had 50,000 pairs. We started selling at 9:45 in the morning, and stopped at 3:12 when the supply ran out. As you might expect, there were customers still on line who were disappointed.

To them we want to say that we're terribly sorry. As the world's largest store, we have proportionately large shipments of nylons—but we have, by far, so many more customers than any other store that it's impossible to supply more

NOTABLE

> When America entered the war, scrap metal became a valuable commodity for the U.S. military. DuPont sacrificed a large portion of its own history by donating 28 iron rolling-mill wheels to the war effort. The original Hagley powder mills had closed down in 1921, idling the 30 massive wheels that used to crush and mix the powder ingredients. Weighing 7 tons each and standing 6 feet in diameter, the 20-inch-wide wheels sat for 20 years as mute monuments to more than a century of DuPont powder-making on the Brandywine. But in 1942 a salvage crew hauled them by tractor from the stone mills to be broken up. "Seems like a shame to do it," said welder Jim Ogden, who used an acetylene torch to separate the paired wheels. "It's like tearing down tombstones." Ironically, small explosive charges packed in mud proved the most efficient means of reducing the iron behemoths to manageable pieces for carting away. Fourteen of the mills' 10-ton iron base plates on which pairs of wheels had ground the powder also went off for recycling. The company kept only two of the wheels for posterity.

> The casual sweater look started with Lana Turner in 1939, but it took DuPont's Orlon® to make the sweater fashion revolution spread across college campuses in the 1950s. Orlon® was mothproof, was much lighter than its natural counterpart, wool, and could be easily washed and worn again. Wool, on the other hand, required careful washing, blocking and pressing. In addition, Orlon® could be dyed to a pure white or bright colors, unlike the more muted tones of dyed wool. The ubiquitous sweater sets in a rainbow of colors, worn with a string of pearls, simply would not have been possible without this new synthetic fabric. Orlon® helped transform sweaters from cold-weather specialty items to year-round wardrobe staples. Whereas Orlon® had only 5 percent of the sweater market in 1953, its share was 50 percent and growing by 1960, with more than 100 million sweaters sold each year.

> During World War II Americans learned to live with shortages of a variety of consumer goods, including the nylon that DuPont diverted from stockings to parachutes. Officials from the government's rationing program reminded consumers that 2,300 pairs of nylon stockings would make a single parachute, whereas a single tin can a week from every American family would make 38 Liberty ships. A shortage of leather hides in 1941–1942, along with greater military demand for boots and shoes, led to shoe rationing for American civilians in February 1943. In 1941, the year that ended with the attack on Pearl Harbor, Americans had purchased an average of 3.5 pairs of shoes each. By 1944 that number was down to two pairs. Still, they fared far better than the heavily besieged British, who were down to one pair a year and who endured severe food rationing well into the postwar years.

> The elements uranium and plutonium were the focus of the Manhattan Project. The nuclei of their atoms contain large numbers of protons and neutrons — 238 for uranium and 239 for plutonium. If an element gains or loses neutrons but keeps all its protons, it forms what is known as an *isotope* of that element. In the 1930s physicists discovered that an isotope of uranium, U-235, was highly fissionable. That is, it easily split into two lighter elements, releasing energy and neutrons in the process. U-235 was also a rare isotope, occurring naturally in the ratio of only one in every 140 uranium atoms. The challenge for Manhattan Project scientists was to separate the U-235 isotope from its parent U-238 and concentrate it for use in a bomb.

Researchers tried three methods to separate U-235 from U-238, but the most successful was *gaseous diffusion*, carried out at the Oak Ridge, Tennessee, facility. Gaseous diffusion pumped uranium hexafluoride gas through a succession of semiporous filters called a cascade. Because the gas molecules containing U-238 were larger than those using U-235, more of the latter passed through the tiny holes of each cascade barrier. At the end, concentrated U-235 uranium hexafluoride was separated into its components, leaving *enriched* uranium and a waste product.

Soon after the discovery of the special properties of U-235, physicists noticed that some U-238 atoms absorbed the neutrons that U-235 emitted when it spontaneously split. Those U-238 atoms then formed the unstable isotope U-239, which sought greater stability by switching some of its neutrons for protons. This created minuscule amounts of neptunium and an entirely new element, plutonium, or Pu-239. Plutonium was also highly fissionable and therefore suitable for an atom bomb.

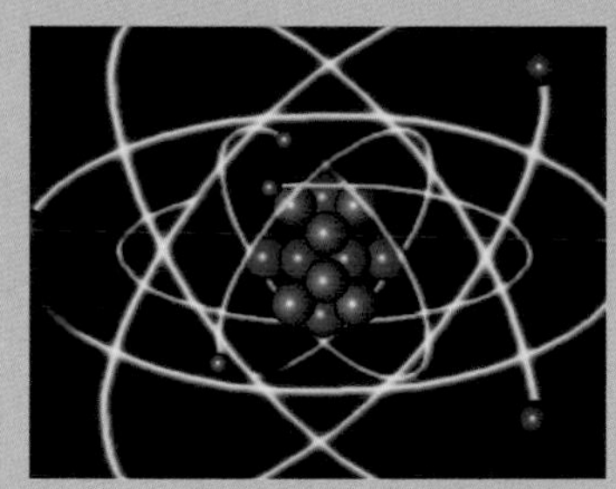

Plutonium was produced in large quantities through bombardment of U-238 with neutrons from U-235 inside a nuclear reactor. However, like U-235, plutonium had to be separated from the larger mass of U-238 and other byproducts of the irradiation process. The Hanford plant in Washington first created plutonium, then separated it from uranium. The Manhattan Project pursued both elements as fissionable cores for nuclear bombs, though only the plutonium bomb was tested at the Trinity site at Alamagordo, New Mexico. A U-235 bomb was dropped on Hiroshima, a plutonium bomb on Nagasaki.

> After its experience in the years following World War I, DuPont feared a revival of war profiteering charges from its work for the military in World War II. To help defuse any such charges should they arise, however, DuPont insisted on a token fee of $1 above costs for all its Manhattan Project work. The fee cemented the formal, contractual nature of DuPont's work for the government, and the minimal fee guaranteed that DuPont would have no capital investment in war-related facilities when peace finally came. When the war ended before the official termination date on DuPont's contract with the government, vigilant federal auditors asked the company to refund 33 cents of the $1 that the government had paid it. Amused DuPont accountants promptly forwarded a check.

Because photographers were not present at the moment of discovery, DuPont reenacted famous scientific breakthroughs. Below, Paul Morgan, a pioneer polymer researcher, demonstrates making nylon in the lab. At Jackson Laboratory, below right, Roy Plunkett (right) reenacts the 1938 discovery of the fluorocarbon polymers that led to Teflon®. This moment also is captured in Plunkett's original lab book notation, opposite.

Experimental Station, DuPont researchers Emerson Wittbecker and Paul Morgan explored a new way to make complex organic substances called polymers at room temperature by suspending their components in two unmixable liquids, much like floating oil on water. At the point where the two liquids met, "interfacial polymerization" frequently occurred. The discovery led not only to a new way to make nylon but also to nylon-based papers, fire-resistant Nomex® fibers and papers, and Lycra® brand elastane in the 1960s. It also encouraged further experimentation with emulsion and spinning techniques, which produced materials such as Tyvek® HomeWrap.

The booming 1950s American economy, it should be remembered, left about a quarter of Americans behind. But for the majority, whose rate of automobile ownership more than doubled in the 10 years after the war; whose TVs and appliances tripled electricity consumption in the 1950s; and whose well-stocked refrigerators in new, suburban homes owed much to the efficiencies of the postwar "agricultural revolution," DuPont's products became an intimate part of a more affluent life. "Better Things For Better Living" had originated in Depression-era hopefulness but was fully realized in the expansive prosperity of the 1950s.

Ironically, Textile Fibers soon faced the same problem that had vexed DuPont's Paint Department 30 years earlier. Rayon, nylon, and Orlon® salesmen were competing against one another, confusing customers with their conflicting claims and undermining one another in the process. The solution was similar to the one chosen in the early 1920s. The various sales divisions were organized centrally as a single, functional division in the Textile Fibers Department, with salesmen promoting all of the department's products instead of just one.

Salesmen in the Polychemicals Department had a different problem with a remarkable but expensive new plastic: it was so unlike any other that no one knew quite what to do with it. Twenty-six-year-old DuPont chemist Roy Plunkett had discovered it one morning in April 1938, when he came to work at the Jackson Lab to continue research on a new refrigerant. He opened one of the 50 pressurized cylinders of tetrafluoroethylene gas he had left on dry ice overnight, but to his surprise nothing came out. Plunkett turned the cylinder upside down and shook it. A small amount of whitish powder wafted to the floor. Plunkett and his assistant then cut the cylinder open and discovered that the TFE had polymerized spontaneously overnight.

Plunkett's first reaction was familiarly human. "Well," he grumbled, "we're going to have to start all over."[17] But his chemist's instincts quickly took over, and he carried the powder to his bench in the lab to take its measure. Immediately he noticed a highly unusual feature. The polymerized TFE would not react or dissolve in any medium, even the harshest acid. After Plunkett repeated the experiment to make sure the polymer could be reliably produced, the substance was packed off to the Plastics Department for further exploration. The results were plentiful. During World War II the material proved useful for the Manhattan Project, in which gaskets made of TFE successfully resisted corrosive acids created in U-235 production. Its temperature resistance made TFE an ideal lining material for liquid fuel tanks, and its superior insulating qualities allowed the use of TFE tape just 0.002-inch thick to wrap the radar wiring on night bombers.[18]

But Plunkett's odd polymer had drawbacks in a civilian market. It was not easily molded, and its extraordinary slipperiness made it difficult to bond with or stick to surfaces. It was also very expensive. The U.S. Gasket Company in Camden, New Jersey, kept the small amounts it ordered from DuPont in a bank vault downtown, where its technicians had to make recorded withdrawals to take back to the plant for processing. DuPont chemists solved the bonding problem by applying non-stick Teflon® in layers, just like paint. The first layer is a primer —

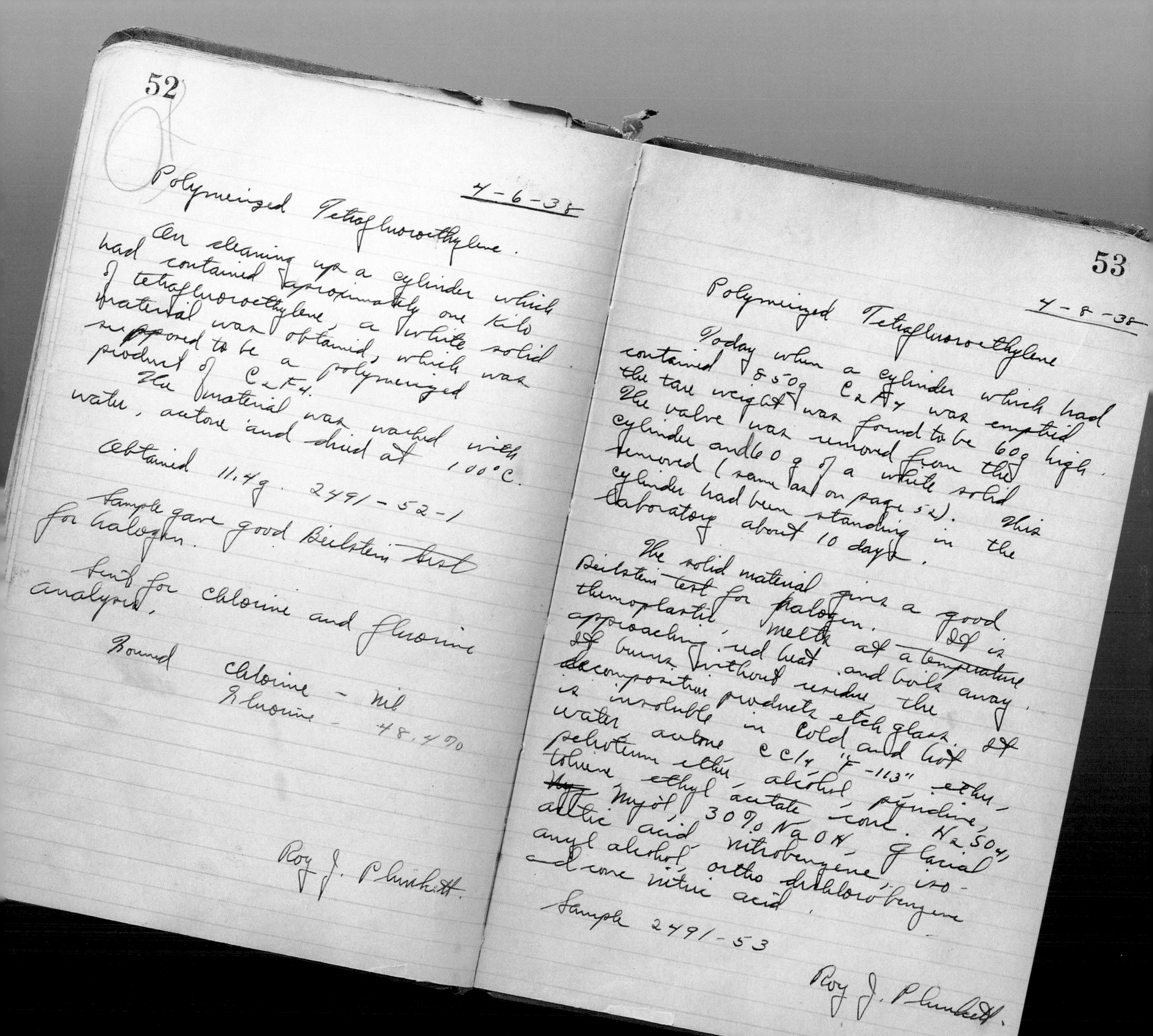

52

Polymerized Tetrafluoroethylene. 4-6-38

On cleaning up a cylinder which had contained approximately one kilo of tetrafluoroethylene, a white solid material was obtained, which was supposed to be a polymerized product of C_2F_4.

The material was washed with water, acetone and dried at 100°C.

Obtained 11.4 g. 2491-52-1

Sample gave good Beilstein test for halogen.

Sent for chlorine and fluorine analysis.

Found Chlorine - nil
Fluorine - 48.4%

Roy J. Plunkett.

53

Polymerized Tetrafluoroethylene 4-8-38

Today when a cylinder which had contained 850 g C_2F_4 was emptied the tare weight was found to be 60 g high. The valve was removed from the cylinder and 60 g of a white solid removed (same as on page 52). This cylinder had been standing in the laboratory about 10 days.

The solid material gives a good Beilstein test for halogen. It is thermoplastic, melts at a temperature approaching red heat and boils away. It burns without residue, the decomposition products etch glass. It is insoluble in cold and hot water, acetone, CCl_4, "F-113", ether, petroleum ether, alcohol, pyridine, toluene, ethyl acetate, conc. H_2SO_4, Nujol, 30% NaOH, glacial acetic acid, nitrobenzene, iso-amyl alcohol, ortho dichlorobenzene and conc nitric acid.

Sample 2491-53

Roy J. Plunkett.

PACKAGING EQUIPMENT PARTS COATED WITH
DU PONT "TEFLON" POLYTETRAFLUOROETHYLENE
"TEFLON"
WITHSTANDS HEAT-SEALING TEMPERATURES
KEEPS PLASTIC PACKAGING MATERIALS FROM STICKING TO MACHINES
SPEEDS PACKAGING
PARTS COATED BY

and it's the chemistry in the primer that makes it adhere to the metal surface of a pan. DuPont made little profit from this plastic until a marketing campaign, much like the one that had worked well for neoprene in the early 1930s, tripled its sales between 1954 and 1960, and a subsequent line of nonstick cookware made Teflon® a household word.[19] Teflon® can now be found in hundreds of other applications from wire and cable insulation to singular configurations of tubing for the manufacture of hard-to-produce pharmaceuticals.

On July 18, 1952, DuPont celebrated two notable achievements even as its legal team, including former U.S. Justice Department attorney Irving Shapiro, prepared for the opening of the GM antitrust trial in November. The first was the company's 150th birthday, an elaborate affair on the site of the original powder mills outside Wilmington that was broadcast to DuPont plants across the country. The second was DuPont's setting of a world safety record for industrial construction — one million man-hours worked without a lost-time injury — at the Savannah River Plant in Aiken, South Carolina, a plutonium production facility under the direction of the new Atomic Energy Commission. World War II, as it turned out, had made some waves as it sank into memory. Winston Churchill's 1946 "Iron Curtain" speech in Fulton, Missouri, provided a lasting image for the profound division between the Soviet Union and communist East Europe, on the one hand, and America and other democracies on the other, that came to be known as the Cold War.

The Soviet Union's successful detonation of a nuclear bomb in September 1949, followed by communist takeover of China the next month and China's entry into the Korean conflict in 1950, created a nervous counterpoint to placid daily life in postwar America. The "mutual assured destruction," or "MAD," doctrine of nuclear deterrence, along with the Truman Doctrine calling for militarized containment of communism around the globe, kept the imminent possibility of apocalyptic war hovering over the nation. In October 1950, at the personal request of President Truman, DuPont agreed to construct and operate the Savannah River plant under the same conditions, and for many of the same reasons, that had governed its participation in the Manhattan Project. DuPont's success at Hanford made it captive to Truman's national security appeal. At the same time, Hanford gave the company a degree of confidence in its ability to construct the Savannah River plant that it had not possessed eight years earlier.

A speech by Crawford Greenewalt at the company's sesquicentennial celebration at the old powder mills informed listeners that in his mind, communism and nuclear war were not the only threats to freedom. So were high taxes and excessive government power. Invoking a mythic sort of individual freedom enjoyed by the company's founder long ago, Greenewalt saw no reason why it should not burn as brightly in the present as it once had for Eleuthère Irénée du Pont. Greenewalt's introduction of edgy ideology into such an amiable event exposed his frustration over the excess profits tax, re-invoked for the Korean War, that subtracted $370 million from DuPont's profits in 1951 — 77 percent more than with the 1949 rates. The speech also reflected his long-standing frustration with federal antitrust law. Apparently, Greenewalt said in June 1949, the government regarded "bigness" as bad anywhere but in the government itself.[20]

Greenewalt's testimony before the House Special Subcommittee on Study of Monopoly Power in November 1949 prompted a statement from its chairman, Representative Emmanuel Celler of New York, that must have deeply offended the DuPont president. Big business, Celler suggested, could actually be paving the way for the triumph of impersonal domination over the individual that many people associated with socialism. In 1950 Congress passed the Celler-Kefauver Act, which stated that prosecutors could consider simple market share evidence of anticompetitive practice, thereby enacting into law what many regarded as merely an uninformed bias: business bigness was bad. Years later

TOP: The celebration of DuPont's 150th anniversary featured a pageant on the banks of the Brandywine with costumed actors as well as the unveiling of a marker on the grounds of the first powder mill.

MIDDLE LEFT: As DuPont president, Crawford Greenewalt was a frequent visitor to company facilities, including this stop in Belle, West Virginia.

BACKGROUND: Plutonium produced in Savannah River Plant reactors was used primarily for national defense but also as fuel in power plants as well as in medical diagnosis and research.

MIDDLE RIGHT: Two rods — one of Teflon® (right) and the second of another resin — are dipped in hot sulfuric acid to demonstrate the properties of Teflon®. The rod of Teflon® is unaffected by the acid, while the other resin is charred and deteriorated.

BOTTOM: Although Teflon® was first identified in 1938, it took more than a decade to develop civilian uses for the material, such as non-stick cookware.

Greenewalt told a meeting of antitrust lawyers in Chicago that he was an authority on antitrust law "only to the extent that an unfortunate pedestrian may be an authority on taxicabs."[21]

Antitrust was more than an arena for philosophical sparring. It shaped DuPont's business strategies through the 1950s and beyond. The Executive Committee avoided acquisitions for fear of triggering further antitrust lawsuits, yet the $90 million it fed annually into research covered less ground with every passing year as laboratory costs kept rising. Some committee members, like David Dawson, saw opportunities in Europe after the creation of the European Economic Community, or Common Market, in 1958, but few of Dawson's colleagues shared his concern about DuPont's maturing product lines in the increasingly competitive chemical industry.

Despite the Cold War's insistent reminders of international tension, domestic prosperity and technological progress dominated American life in the postwar years. The number of scientists in America increased by an astounding 930 percent between 1930 and 1964. After World War II DuPont employed many of them — 2,300 by 1958 — to deliver on science's promise to an affluent society. The arrangement seemed too good to alter, its assumptions too fundamental to question, its rewards too consistent for anyone to call too loudly for change.

DuPont's red oval was a familiar landmark on America's progress toward a better future. But as time muted the memory of World War II's sacrifices, and the Cold War's nuclear Sword of Damocles hung dangerously over the rush of daily life, many people began to wonder whether the country had lost its way. Some at DuPont also looked past the company's enormous postwar prestige and confidence and wondered about the future. DuPont faced a maturing industry, rising costs, and new sources of competition. New strategies would be required, but so would a new consensus within the organization. Like the accelerating "click-click" from Fermi's pile, the questioning of the 1960s and 1970s gradually mushroomed into a clamorous pressure for change.

PONT
®

CHAPTER 7

REACHING OUTWARD, LOOKING INWARD

RIGHT: DuPont's facility in Londonderry, North Ireland, opened in 1960 to produce neoprene primarily for United Kingdom and European markets.

From the rubble of World War II — seen here (far right) in Lunebach, Germany — emerged a new European economy spurred by the Marshall Plan, named for Secretary of State George C. Marshall. Meeting (center right) to discuss the European Recovery Program of 1948 were (l to r): President Harry Truman; Marshall; Marshall Plan administrator Paul Hoffman; and Secretary of Commerce Averell Harriman.

NEAR RIGHT: As a member of the Executive Committee, Samuel Lenher pushed DuPont's international expansion.

BOTTOM: DuPont's Executive Committee of 1957, shown around a table in the shape of the DuPont oval.

The 1953 court-ordered breakup of DuPont's shared businesses with ICI, Ltd., in Canada, Brazil and Argentina set these two chemical industry giants in fresh competition for international sales. In 1956 a team from DuPont's Organic Chemicals Department invaded ICI's home territory in the United Kingdom, searching for a factory site. The ancient city of Londonderry in Northern Ireland was chosen as the home for a new neoprene plant. The DuPont Company, Ltd., chartered in London in 1956, opened the facility in 1960, just two years after the founding of the European Economic Community. The Common Market, as the EEC came to be known, lowered tariffs and other barriers among participating European countries, making overseas investment attractive to DuPont and easing the entry of the Londonderry subsidiary's neoprene into a flourishing new market.

DuPont's Londonderry investment was evidence of a wider U.S. confidence in a rejuvenated European economy. Europe had largely recovered from the war's devastation with help from the United States' Marshall Plan. DuPont, along with many other U.S. businesses, expanded its international operations around the globe in the late 1950s, not because international competition was weak but because the company now had to meet sharp challenges from new, sophisticated rivals. The United States had helped promote this healthy rivalry not only through the Marshall Plan but also by spearheading the 1944 Bretton Woods agreements that established the World Bank and the International Monetary Fund and led to the 1947 General Agreement on Tariffs and Trade (GATT). Nations participating in periodic GATT negotiations, or "rounds," agreed to lower tariffs and other traditional barriers to free trade.

Although DuPont enjoyed great prestige and success as its trademark oval logo spread around the world in the early 1960s, the company's top executives were fully aware of encroaching problems of product maturity, declining prices and increased competition at home and abroad. Continued research and manufacturing improvements such as those achieved in nylon production by the Textile Fibers and Engineering Departments helped keep DuPont's businesses profitable. In the new world of international competition, however, innovation ensured only the skillful maintenance of a narrow business advantage, not the security of overwhelming superiority that DuPont once had possessed.

With the encouragement of David Dawson, Samuel Lenher, and other members of the Executive Committee, Crawford Greenewalt devised two strategies to meet this change. First, DuPont enlarged its international business; second, it relied on its legendary research strength to bring new products to market, not on the scale of nylon's giant stride but in smaller, more deliberate steps toward the consumer. This strategy involved making not only basic materials but also retail products — recording tape, antifreeze and pharmaceuticals. Greenewalt likened R&D to a rain barrel: as old products drained out the bottom, research would pour new

BACKGROUND: **DuPont opened sales and distribution facilities throughout the world in the 1950s.**

TOP LEFT: **By 1958, 2,400 scientists carried on research at DuPont.**

TOP RIGHT: **DuPont Canada's products include synthetic fibers, polymer resins, packaging films, automotive finishes, crop protection products and industrial chemicals.**

MIDDLE LEFT: **Mylar® polyester film, commercialized in 1952, replaced cellophane as the major product of the DuPont Film Department.**

MIDDLE RIGHT: **Nylon de Mexico S.A. was formed in 1974.**

BOTTOM LEFT: **Lammot du Pont was the ninth, and last, DuPont president descended directly from the company's founder.**

BOTTOM RIGHT: **Scientist Paul Morgan demonstrated the interfacial polymerization process — "the nylon rope trick" — which was fundamental to the development of what became Nomex® brand fiber.**

ones in at the top, keeping the contents fresh.

In 1958 DuPont reorganized its Foreign Relations Department as the International Department, under W. Sam Carpenter III, to coordinate the company's burgeoning businesses in Europe, Asia, Canada, Mexico and South America. Two years later, toward the end of his long tenure at DuPont's helm, Greenewalt encouraged all the industrial departments, as well as the Development Department, to direct their research toward new products and new markets, wherever they might be found. Expropriation of a new $1.2 million paint plant in Cuba by a young revolutionary named Fidel Castro in 1960 recalled warnings by former Executive Committee members like Angus Echols about the hazards of "foreign" investment, but this setback did not sway the company from its new course. DuPont purchased the prestigious, century-old German photo film company Adox Fotowerke in 1962 and formed four jointly owned companies in Japan between 1960 and 1964.

DuPont's new operations in Europe and Asia were well under way when Lammot du Pont Copeland, a great-great-grandson of the company founder, succeeded Greenewalt as company president in August 1962. With a chemistry degree from Harvard, Copeland had started working for DuPont in 1929, was laid off briefly during the Depression, then returned to work at the company's Fairfield, Connecticut, Fabrics and Finishes plant for five years. He next joined the Development Department at Wilmington headquarters, was named to the Board of Directors in 1942 when his father, Charles Copeland, retired, and became chair of the company's Finance Committee in 1954. Like his uncle Pierre, Copeland was a shy man; like his uncle Alfred, he was hard of hearing and sometimes had trouble following discussions at meetings. His personal lifestyle reflected his wealth — cattle breeding in Maryland, duck hunting on the Chesapeake Bay and salmon fishing in Scotland; a French chef, gourmet dining and fine wines at home. But Copeland's office at DuPont headquarters in Wilmington was unpretentious. He put neither his name nor his title on the door and drove himself to work every morning in his low-budget Chevrolet Corvair.

As the company's largest single stockholder, Copeland had good reason to celebrate the company's $452 million in earnings in 1962 — the best showing in the company's 160-year history. But DuPont's costs, especially its research and development outlays, were steadily rising at the same time that competition was driving down the prices the company could charge for its products. Caught in what economists call a "cost-price squeeze," DuPont followed the plan that the Executive Committee under Greenewalt, then Copeland, had established to keep this slow squeeze from shrinking profit margins. This dual strategy of expansion and innovation was initially very successful. DuPont's $2.4 billion in sales in 1962 was the largest total sales of any chemical company in the world that year and one-third higher than that of its nearest rival, ICI. International sales of $388 million were twice the 1957 level; exports from U.S. plants roughly equaled sales by new DuPont subsidiaries in Japan, Mexico, Canada, Colombia, Peru, Argentina and Western Europe.

DuPont marked nylon's 25th anniversary in 1963 by spending a record $370 million to upgrade and expand nylon production plants. The Textile Fibers Department opened a pilot plant in Richmond, Virginia, for a new, heat-resistant fiber called Nomex® that the company believed would be useful to military and aerospace customers, while the Fabrics and Finishes Department announced that its long search for a satisfactory synthetic leather had at last ended. In the early 1950s, three departments — Textile Fibers, Fabrics and Finishes, and Film — had competed to develop a porous, moisture vapor permeable sheet material that would "breathe" like leather but last longer with less maintenance. In 1955, deciding that Fabrics and Finishes, maker of nonporous Fabrikoid, was most likely to succeed, the Executive Committee authorized the department to carry the research forward on its own.

DuPont Corporate Centre

TOP: **In the 1960s, DuPont's Color Council experimented with new uses of color on packages to attract consumers. Council chairman was Domenico Mortellito, a prominent sculptor, muralist and painter.**

BOTTOM: **With new dye technologies, DuPont added color throughout its product line, including wigs made of Tynex® nylon filament, more commonly seen as brush filaments.**

OPPOSITE: **To demonstrate the bounty a typical American family enjoyed over the course of one year, DuPont photographer Alex Henderson carefully posed DuPont employee Steve Czekalinski with his wife and two boys amid 669 bottles of milk, 578 pounds of meat, 131 dozen fresh eggs, 440 pounds of fresh fruit as well as coffee, cereal, flour and much more. The picture appeared in the November 1951 issue of *Better Living* magazine and was carried later in *Life* magazine. The picture subsequently was updated twice.**

Over the next several years DuPont technicians worked closely with shoe manufacturers to ready their new "poromeric" material for final production. As their counterparts had done a generation earlier with neoprene and nylon, these technicians guarded their product against industrial spies, diligently sweeping up stray scraps of synthetic leather from factory floors to keep them from straying into competitors' hands. Then postal workers, police officers, and DuPont's own executives and sales personnel field-tested a total of 15,000 pairs of new synthetic-leather shoes for durability and comfort. Only 8 percent of wearers complained about the shoes — one-third the number of complaints lodged about nonporous vinyl shoes and slightly more than double the number for leather shoes. Encouraged by these results, the Executive Committee decided in October 1962 to move from pilot plant to full, commercial production. Two years later DuPont's venerable Old Hickory plant in Tennessee started making Corfam® with an ambitious goal of 30 million square feet a year, and its Malines plant in Belgium prepared to dye and finish Corfam® for European customers. With a workforce of more than 100,000 "DuPonters" behind him, Copeland, who wore his own pair of Corfam® shoes faithfully every day, appeared to be marching the company toward a bright future.

It was not a lonely march. Inspired in part by the success of DuPont's discoveries, nearly half of America's 500 largest manufacturers, including such diverse businesses as General Foods and Grace Shipping Lines, had joined the "gold rush" into chemicals after World War II. The chemical industry grew twice as fast as U.S. industry as a whole in those years. Although DuPont remained the world's leading chemical producer in the mid-1960s, output was spread so widely that its sales accounted for only 7.5 percent of the industry total.[1] Many other companies in America and around the world had bright Ph.D. scientists and research labs of their own, ready to challenge the industry leader with new as well as what DuPont's Edgar Woolard later called "me too" products.

WHEATIES
SUNNYFIELD CORN FLAKES
CORN FLAKES
Extra-Big-Pak
Beech-Nut COFFEE
CRISCO
ALBRO SAUERKRAUT
Sunnyfield FAMILY FLOUR ENRICHED
JACK FROST CANE SUGAR
SAN JOAQUIN VALLEY
FULL OF SUGAR
PYRAMID ORCHARDS APPLES
Eggs
EGGS HANDLE WITH CARE
POTATOES
PINK LADY TOMATOES
U.S. No.1 ONIONS
VANILLA
MILK

NOTABLE

> Postwar Europe eventually provided a great market for expanding American companies like DuPont. But initially the region's needs were too great for private business alone to meet. The war had left Europe, in Winston Churchill's words, "a rubble heap, a charnel house, a breeding ground of pestilence and hate." In the summer of 1947, United States Secretary of State George C. Marshall advocated a program of massive economic aid to rebuild Europe. The unusually harsh winter of 1947–1948 brought starvation to many Europeans, but it helped the soft-spoken former U.S. Army general overcome isolationist opposition to his aid plan. When the Soviet Union invaded Czechoslovakia in February 1948, the United States finally gave full support to the Marshall Plan as part of its strategy to contain the spread of communism. Between 1948 and 1951 the Marshall Plan funneled $13 billion into European recovery, which soon rewarded the United States with new markets. Marshall's efforts earned him the Nobel Peace Prize in 1953.

> DuPont established a major administrative subsidiary, DuPont de Nemours International S.A. in 1961 in Geneva, Switzerland, to act as a central sales and export company for European operations. The subsidiary also supervised DuPont's sales offices in 78 countries worldwide.

> Building on prior Dacron® research, DuPont chemists in 1952 developed Mylar®, an exceptionally strong polyester film for use in electrical, electronics, magnetic audio and video recording, imaging and graphics, and packaging applications.

> By the late 1950s, DuPont manufacturing expanded rapidly outside the United States. Production began in 1957 of Freon® in Argentina and Brazil. A finishes plant opened the same year in Venezuela. DuPont formed a subsidiary with Mexican investors in 1958 for the production of TiO_2 in Altamira, Mexico. An automotive paint plant opened in Mechelen, Belgium, in 1959, the company's first manufacturing plant on continental Europe.

> The consumer revolution of the 1950s and 1960s brought bright color options into the bathroom, the kitchen, the automobile and the closet. DuPont's new engineering plastics, coatings and synthetic fabrics, such as Delrin®, Alathon® and Tynex®, appeared in power tool casings, cups, soap dishes, wigs, bathing suits, milk cartons and tubes for everything from toothpaste to cake icing. In the 1960s the DuPont Color Council helped design and market these new colorful items. Council chairman and well-known artist Domenico Mortellito reveled in the new design possibilities. Mortellito, famous for his 1930s murals in zoos, post offices and World's Fair pavilions, pioneered the use of synthetic materials in fine arts. His designs for DuPont exhibits at airports and other public venues used color extensively to attract attention. Though DuPont discontinued its dye business in the 1980s, the company continued to support research into electronic color-measuring instruments and color physics. Today, color managers in several DuPont departments serve America's continuing love of colorful merchandise.

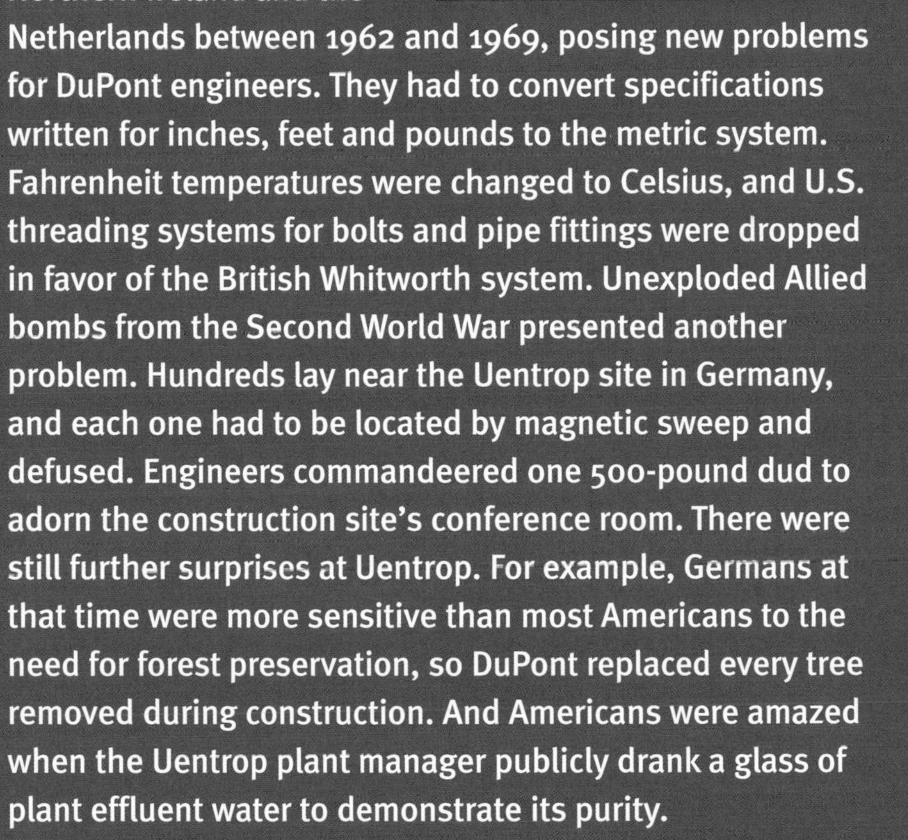

> "DuPont fiber plants seemed to spring up like weeds in Europe throughout the 1960s," recalled DuPont employee D. F. Holmes. Nine such plants opened in Germany, Northern Ireland and the Netherlands between 1962 and 1969, posing new problems for DuPont engineers. They had to convert specifications written for inches, feet and pounds to the metric system. Fahrenheit temperatures were changed to Celsius, and U.S. threading systems for bolts and pipe fittings were dropped in favor of the British Whitworth system. Unexploded Allied bombs from the Second World War presented another problem. Hundreds lay near the Uentrop site in Germany, and each one had to be located by magnetic sweep and defused. Engineers commandeered one 500-pound dud to adorn the construction site's conference room. There were still further surprises at Uentrop. For example, Germans at that time were more sensitive than most Americans to the need for forest preservation, so DuPont replaced every tree removed during construction. And Americans were amazed when the Uentrop plant manager publicly drank a glass of plant effluent water to demonstrate its purity.

> A team of DuPont scientists led by research director W. Frank Gresham in 1954 synthesized two new substances, linear polyethylene and polypropylene. Because of the scientific, technological and commercial breakthroughs achieved by the development of these dual products, the film, bottling, textile fiber, monofilament and molding industries were revolutionized almost overnight.

DUPONT USED MAGAZINES, TELEVISION AND RADIO FOR ADVERTISING IN THE 1950s, AS WELL AS EXHIBITS AT FAIRS, AIRPORTS AND OTHER PUBLIC PLACES. IN THE 1950s AND 1960s, DUPONT INTENSIFIED ITS CAMPAIGN TO MAKE END-USE PRODUCTS FOR DIRECT SALE TO CONSUMERS. BRAND NAME RECOGNITION WAS A KEY ELEMENT OF THE COMPANY'S DIRECT-TO-CONSUMER MARKETING STRATEGY. BY 1962, THERE WERE 225 PRODUCTS WEARING THE OVAL, INCLUDING CAR CARE PRODUCTS, PAINTS, SPONGES AND COMBS. DUPONT MADE OVER 600 COLORS OF INTERIOR WALL PAINT IN 1959. MEMBERS OF THE DUPONT ADVERTISING DEPARTMENT (above) CONSIDERED HUNDREDS OF NAMES FOR NEW PRODUCTS. E. R. MULLIN LED A TEAM THAT SELECTED "SLIPSPRAY" FOR A NEW, DRY LUBRICANT.

GREAT
REFLECTIONS
CAR WAX
Reflects the brightest beauty of your car's finish

RAIN
DANCE
PASTE CAR WAX

DuPont
family
pack
98¢
DuPont
Value Pack
VARNISH
FLAT
DULUX
DUCO
DULUX
Super-white
LUCITE
House Paint
HOME OF THE
WorkSkipper PAINTS
DU PONT
ORIGINAL
DuPont
PURE CELLULOSE
Sponge
Absorbs 20 times its weight in water!
DU PONT
YEAR ROUND
ZEREX
SUMMER COOLANT
ANTI-FREEZE
GAS
Booster
CONCENTRATE
TREATS
20
GALLONS

WORLD'S
FAIR
WORLD'S
FAIR
WORLD'S
FAIR
WORLD'S
FAIR
DUPONT
WORLD OF CHEMISTRY
Feria Mundial de Nueva York 1964-65
Fiera Mondiale di

By mid-decade, DuPont's customary three-to-five-year lead time with new products had nearly evaporated. From 1964 through 1967, sales and operating investment climbed steadily higher while prices and net earnings fell. DuPont was running faster just to stay ahead — expending greater efforts for smaller rewards — though this inflationary dynamic was momentarily overshadowed by the larger prosperity that DuPont's technical innovations had produced.

DuPont's success partly reflected the unprecedented rise in productivity and living standards that characterized American society in the two postwar decades. After John F. Kennedy was assassinated in November 1963, President Lyndon Johnson capitalized on the robust economy and the liberal social mandate of the times to promulgate a massive civil rights and social improvement program nicknamed the Great Society. At the same time, he followed the advice of Kennedy's advisors in pursuing and escalating the undeclared war against communism in Vietnam. Could America afford to do both? "Hell," Johnson drawled, "we're the richest country in the world, the most powerful. We can do it all."[2] Johnson's bravado rested squarely, or so it seemed, on the underlying assumptions of the "new economics" of postwar affluence, which judged the economy to be so strong, so fundamentally stable, that it was virtually recession-proof. All that was needed was an occasional Keynesian spill from brimming federal coffers to keep the fertile soil moist. This came in 1964 when Congress passed a $10 billion tax cut to keep the good times rolling.

Fifty-one million people enjoyed a hallmark of the good times: the New York World's Fair, which opened on April 22, 1964. A great number of them visited DuPont's "Wonderful World of Chemistry" exhibit before the fair closed in 1965. Many people entering the company's two-story, circular pavilion with its necklace of plastic globes may not have noticed the 48 DuPont materials used in its construction — Tedlar® roof covering, Delrin® door handles, Mylar® stage curtains, and nylon carpeting, to name a few — but the

DuPont, increasingly an international company, printed brochures in six languages for the 1964 World's Fair. The carousel-shaped DuPont pavilion incorporated 48 DuPont products in its construction and appointments. To remind visitors that DuPont has always been a science company, molecular structures were featured in the design. Women competing for "prettiest legs at the fair" wore hose of Cantrece® nylon.

The Goodyear blimp *Columbia* was constructed in 1963 with Dacron® polyester coated with neoprene.

Milk cartons — dubbed "A Square Package with a Hip Pitch" — were coated with Alathon® polyethylene.

Plastic squeeze tubes — many made with DuPont resins — were the 1960s rage.

Industrial designers of 1960 were using materials that did not exist 20 years earlier. These included Delrin®, Zytel®, Alathon®, Teflon® and Lucite®.

A lacrosse stick of Adiprene® urethane rubber (left) offered more durability than wooden sticks.

Brooms of long-wearing nylon filament clean 800,000 miles of New York City street each year.

exhibit's "Parade of Products," "Plastics Family" and "Chemical Magic" shows provided genial enlightenment.

In the "Wonderful World of Chemistry" exhibits DuPont proudly displayed its scientific expertise and reminded its guests of the company's important role in providing them with "Better things for better living." Alathon® polyethylene resin typified DuPont's often low-key, almost anonymous, contributions to material and consumer culture when it appeared in 1962 as a new milk carton coating. Even nylon, for all its fame as hosiery, also served quietly and with less glamour as cording in half of all the tires produced in the United States in 1963. Surlyn® ionomer resin, introduced in 1964, was another versatile polymer, transparent and tough, that appeared in many forms, from films and packaging to molded plastics in automobiles and recreational equipment, such as golf balls. Teflon®, on the other hand, was more widely recognized by consumers in 1965 because by that time it coated 40 percent of all cookware manufactured in the United States.

Diversification and internationalization were DuPont's key strategies for continued success in the 1960s. But the company's organizational structure sometimes worked against smooth implementation of these strategies. Semi-autonomous industrial departments tended to concentrate on short-term profitability of their tried-and-true products, while the independent Central Research Department leaned in the other direction, pursuing research interests not directly tied to products or profits. The creative tension between business and science had long been an underpinning of DuPont's dynamism, but by the late 1950s Crawford Greenewalt and the Executive Committee had recognized the need for some additional energy, a "booster shot" for the old formula.

In 1960 Greenewalt looked to the Development Department to help spur what the company called its New Venture program. He asked Edwin A. Gee, a chemical engineer and director of sales in the Pigments Department, to join the Development Department and brainstorm some new businesses "beyond what the industrial departments are now doing." Gee, a rare "outsider" who had come to DuPont from the Bureau of Mines and had played key roles in the Pigments Department's titanium and silicon ventures, remembered being "scared to death" by the large responsibilities of the new assignment. He also worried about working largely on his own, outside the boundaries of Central Research and the industrial departments. But Gee set aside his doubts, accepted Greenewalt's challenge, and set off to meet with the director of the Development Department, Henry Ford.

Ford's welcome did little to allay Gee's worries. "I want you to know from the beginning I have no faith in this effort," he told his new assistant director. "If I was going to start it, I wouldn't have picked you. I'm going to delegate to you all of the authority I have, and you can be the biggest success or the biggest failure in the world, and I don't give a damn."[3] Gee and Ford eventually became close friends, but on that first day Gee stayed in his lonely office pondering the task ahead. After a few months he returned to the Executive Committee with eight new ideas, expecting that Greenewalt and the committee might find one of them acceptable. "Do all eight," they said.

Gee was stunned by the committee's response and by the scope of his new mission. DuPont had undertaken such a program in the 1920s, but at that time it had relied as much on acquisitions as on its own research facilities to find new products. Now, after years of bruising antitrust prosecution by the federal government, acquisition seemed a hazardous route to take, yet in-house research and development costs were skyrocketing. As a percentage of earnings, R&D spending at DuPont had doubled between 1946 and 1961 and continued to rise — by a total of 25 percent — between 1962 and 1969. Gee had earned a reputation for harnessing R&D costs in the Pigments Department, but he also realized how difficult it would be to turn research to advantage without the option of acquiring it from outside. Later he emphasized that Greenewalt's rain barrel "simply wasn't a valid strategy

GOODYEAR
GOODYEAR
NEW
ELKO
hand
aid
40

ABC
AMFLITE II
AMF

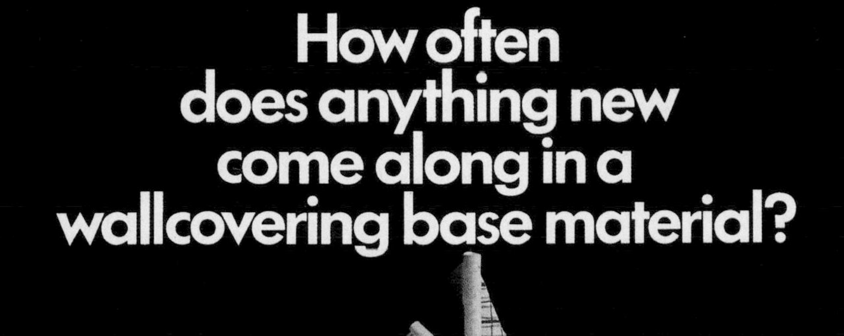

OPPOSITE TOP LEFT: **Lycra® brand elastane was first produced at DuPont's Waynesboro, Virginia, plant.**

OPPOSITE TOP RIGHT: **Corian® solid surfaces came to market in the 1960s from DuPont's "New Venture" program.**

OPPOSITE BOTTOM LEFT: **Coating bowling pins was an early application of tough Surlyn® ionomer resin.**

OPPOSITE BOTTOM RIGHT: **The Automatic Clinical Analyzer enabled laboratories such as this medical center in Caracas, Venezuela, to perform dozens of tests on blood serum and body fluids in about seven minutes.**

DuPont introduced Tyvek® protective material as a wall-covering base in 1969.

anymore because you couldn't bring enough things in at the top of the barrel."[4]

Nevertheless, Gee pressed on with his new projects. Two, in photocopying and cast nylon, soon were abandoned as unlikely to overtake DuPont's strong competitors in those fields. The remaining projects — scientific instruments, medical equipment, heat transfer products such as radiators, building products, magnetic tape, and "embryonic ventures" (seeding small technology companies) — yielded mixed results. As a group they cost $74 million to develop but netted only $30 million in earnings by 1975.[5] DuPont's Automatic Clinical Analyzer for medical diagnostic testing had nearly recouped the company's investment by 1975 and went on to become a very successful product, but the payoff on other New Venture products like Corian® solid surfaces material and Crolyn® magnetic tape took much longer. DuPont's diversification efforts also branched out into molecular biology and prescription drugs. In October 1966 the U.S. Food and Drug Administration approved DuPont's new influenza prophylactic, Symmetrel®, a drug that medical researchers later found useful in treating the symptoms of Parkinson's disease.

All in all, DuPont spent $100 million to launch 41 new products in the 1960s. Many were great successes. Agricultural chemicals, like the fungicide Lannate® that DuPont introduced in 1968, helped drive a continuing agricultural revolution around the globe. Lycra® brand elastane, one of DuPont's most successful products, reached its goal of $50 million in sales by the end of the decade, and new ways of formulating old products also proved profitable. For example, a type of nylon called "bulked continuous filament," or BCF, captured 40 percent of the carpet fiber market by 1966. The Old Hickory plant delivered 3 million square feet of Corfam® to shoe manufacturers in its first year of production. Backed by a $2 million advertising campaign, Corfam® was so well received that two years later, in 1966, production reached 20 million square feet, all of which was quickly sold.

Then the vagaries of the fashion market knocked Corfam® off its stride. The U.S. shoe industry suffered in the mid-1960s from a flood of low-cost leather imports. Corfam® shoes couldn't compete because of the material's high production costs. Corfam® also was difficult to "tool," thereby losing out to leather in the constantly changing women's shoe market that favored a wide variety of ornate designs. Such tailoring was difficult in large-scale, continuous production of sheeted material like Corfam®. But the biggest complaint from consumers was a seemingly irrational one that their shoes wouldn't "stretch" or "breathe." DuPont marketers were befuddled by such feedback, for their data showed that Corfam® did indeed "breathe" and that customers in field tests had been mostly satisfied with the product. Besides, Corfam® shoes weren't meant to stretch. Why should they if they were properly fitted? Ironically, that was the problem. Shoe salesmen were used to fitting shoes on the assumption of some future "give" in the leather. They resisted adopting two separate ways to sell their products, so they often sold Corfam® shoes that were in fact too tight.

Corfam® was at last discontinued. DuPont sold the rights and all the production equipment to a Polish buyer in 1971 after painfully concluding that this "excellent product" would never recoup the company's huge $80 million investment. Years later DuPonters remained protective of their star-crossed product. "After twenty-five years or more, these shoes are my best shoes," boasted the Development Department's Chaplin Tyler. "They've been resoled about six times, and nobody will ever wear them out."[6] Senior Vice President Irving Shapiro, who had monitored the Fabrics and Finishes Department

for the Executive Committee and had actually pushed for ending Corfam® production, agreed. "If Corfam® had been invented ten years later," he said, "I'm confident it would have been a *huge* success."[7]

The New Venture program's middling success and the unexpected demise of Corfam® chipped away at DuPont's self-confidence as much as its bottom line. Though the company's public position highlighted its traditional readiness "to risk much to gain much," as Copeland put it in 1966,[8] the Executive Committee privately had been seeking evaluations from departmental managers as part of a searching self-examination. The appraisers pulled no punches. Textile Fibers' Andrew Buchanan and Lester Sinness had reported as early as 1961 that "a marked lack of confidence in Company leadership has developed due to our slow reactions to changing conditions, insufficient forward planning and excessive confidence in 'fundamental research.'"[9] Three years later Sinness wrote Copeland that DuPont's overall research output was "disgracefully and inexcusably low in proportion to the calibre of men we employ, the facilities we give them, and the amount of money we allow them to spend."[10] Sinness, who had joined the Executive Committee in 1962, claimed that its monthly "chart room" reviews had become ritualistic rather than useful and undermined the committee's intended purpose of focusing on DuPont's long-range goals. Sinness was especially outspoken, but others, including David Dawson and Edwin Gee, agreed.

The Executive Committee had usually made decisions by consensus rather than formal majority vote. Under Greenewalt's forceful direction and persuasiveness, consensus had worked well. But Copeland's less assertive leadership often left the committee bogged down in indecision rather than at risk for the discord that a mere majority vote might generate. The result was an unresolved dilemma at the top levels of management. On the one hand, DuPont had accepted the necessity of international expansion and a new research effort focused on diverse, marketable discoveries — what Gee aptly called "the vigorous exploitation of internally generated inventions."[11] On the other hand, the company could not accept the necessity of radically changing its own organization in order to reach those goals.

DuPont's leadership was certainly not nearsighted. Most key executives had given up as nostalgic any hope for a "new nylon" — a single, high-risk research gamble that would pay off handsomely in long-term market dominance. Yet DuPont's organizational structure almost ensured that the company's New Venture program would be conducted along costly old lines. The quarter-billion dollars the company spent annually on R&D in the mid-1960s exceeded the combined amount spent by three major competitors — Monsanto, Union Carbide and Celanese. DuPont was weighed down by its 30-year-old assumption that profits and science were best connected by putting science on a long leash. At the same time, the managers of its industrial departments, though loath to give up their traditional semi-autonomy, bemoaned an Executive Committee that seemed remote and unable to move the company as a whole in new directions.

Much of the dissension among DuPont's executives in the 1960s concerned the company's expensive commitment to long-term, basic research. Many believed that the new competitive climate in the chemical industry demanded a research shift toward improvements in specific, marketable products rather than open-ended investigations. Frank McGrew, research director in the Polychemicals Department from 1958 to 1964, cited the case of respected scientist Frank Gresham, who held the company record for filing patents. "What staggered me," said McGrew, "was that not a single bloody one of them had ever had any commercial use."[12] McGrew's comments aside, Gresham's research teams discovered linear polyethylene and polypropylene, work for which he received the company's highest honor, the Lavoisier Medal.

Predictably, the tension between basic science and

TOP: According to advertising, stockings of Cantrece® nylon had a "subtle, matte finish for a new kind of natural flattery, a luxurious softness against the skin, and built-in resilience that shapes them to the leg." The model's shoes were made of Corfam® poromeric.

MIDDLE: The Brandywine Building — the third major DuPont office building in Wilmington, Delaware — opened to more than 2,000 employees in the early 1970s.

BOTTOM: The Chambers Works on the New Jersey side of the Delaware Memorial Bridge produces some 500 products, including stain protectants used in Stainmaster® carpets and ingredients for Kevlar® brand fiber.

Americans saw bright possibilities for change in the 1960s, but the sunshine often seemed to disappear in storms of discouragement and division. The assassinations of John F. Kennedy, Martin Luther King Jr. and Robert Kennedy, combined with assaults on civil rights marchers and widespread urban rioting, eroded Americans' confidence in their basic social unity. The lengthy, inconclusive war in Vietnam further polarized society, exhausting both the U.S. Treasury and the energies of the indomitable President Lyndon B. Johnson. Yet despite these wrenching conflicts, Americans in the 1960s extended their respect for individual rights and social justice, and nurtured a newfound appreciation for natural resources. For DuPont, and for all Americans, business would never be the same.

applied science rose whenever earnings dropped. Since the late 1920s DuPont had maintained a strong faith in the ultimate commercial value of basic research. By the 1960s the company's Experimental Station had become a kind of church ministered by scientists who, if they were aware of the collection plate, seemed sheltered from any real fear of the bill collector. Chemists Howard E. Simmons Jr. and Stephanie L. Kwolek each spoke of the 1960s as "a glorious period" and "the golden age" at DuPont. Kwolek recalled "a tremendous amount of independence" in pursuing her research interests. When her research director dropped in at the Experimental Station from his downtown office to ask what she was doing, and why, she remembered, "I used to wonder why the heck he was bothering me."[13] By 1972, DuPont's R&D costs per research employee exceeded the chemical industry's average by 18 percent. Just one year later that gap had grown to 23 percent.

Inflation fueled by steadily rising government spending in the mid-1960s increased the pressure on DuPont to sharpen its focus on reducing costs. Lyndon Johnson's Great Society agenda included Medicare and Medicaid, civil rights and voting reforms, and massive programs to improve education, fight poverty and clean up the environment. The cumulative costs of launching these programs reached nearly $20 billion by the mid-1960s. The Vietnam War cost even more. By 1968 the United States had spent $100 billion in Vietnam, with no end in sight. Federal deficits ballooned from $1.4 billion in 1965 to $25 billion in 1968, when a long-avoided tax surcharge restored some balance to the budget.

In 1966 industry overcapacity and worldwide competition finally caught up with the synthetic fibers market and sent prices plummeting. Wall Street responded accordingly, and the value of DuPont stock dropped 40 percent. In November 1967 Lammot Copeland met with a group of executives and editors from the *Wall Street Journal* to combat rumors that DuPont was, as he put it, an "old, tired company . . . stuck on dead center."[14] But in truth, Copeland's

vision of high-stakes research gambles, combined with the inertia of a deeply entrenched organizational structure, suggested little in the way of hope or new direction for the company. Like the Cold War political consensus of communist containment that was dissolving in Vietnam, the old economic consensus about DuPont's long-range research investments no longer seemed persuasive to investors, or to many of the company's managers.

In December, just one month after Copeland's meeting with the *Wall Street Journal* editors, DuPont's Board of Directors named his successor — Charles Brelsford McCoy, son of a former DuPont vice president. McCoy was a man steeped in the company's ways and traditions. He had started as an equipment operator at the company's Richmond, Virginia, cellophane plant in 1932 when jobs were scarce even for University of Virginia chemistry graduates. Later he earned his master's degree in chemical engineering at MIT and worked his way up through industrial department positions, including four years in DuPont's London office, before being appointed to the Executive Committee in 1961.

Beyond his deep loyalty and his familiarity with the company, McCoy possessed some qualities that were especially timely for DuPont's situation in the late 1960s. He was a shrewd and patient listener who kept his poise in adversity. He could be both flexible and judicious, willing to adjust the company's traditions to the changing temper of the times without yielding to pressures for radical alterations. Incredibly, the events that had swamped Copeland multiplied around McCoy, who became DuPont's 12th president just as one of the most turbulent years in American history was erupting. The assassination of Martin Luther King Jr. triggered a wave of urban riots. The assassination of Robert Kennedy just two months later extinguished the hopes of many for an early end to the Vietnam War and aggravated the widening, acrimonious division among Americans over the purpose of that conflict. The sweeping social unrest of 1968 marked the

The 1968 Executive Committee: (seated, left to right) Samuel Lenher, Lester S. Sinness, Charles B. McCoy, David H. Dawson; (standing) George E. Holbrook, Joseph A. Dallas, Irénée du Pont Jr., Wallace E. Gordon, R. Russell Pippin, and Samuel A. Milliner Jr., secretary to the committee.

Charles B. McCoy led the company from 1967 to 1973, a turbulent, challenging time for the nation as well as the company.

DUPONT SCIENTIST STEPHANIE L. KWOLEK DEVELOPED THE FIRST LIQUID CRYSTAL POLYMER THAT PROVIDED THE BASIS FOR KEVLAR® BRAND FIBER. SHE WAS AWARDED THE NATIONAL MEDAL OF TECHNOLOGY IN 1996 FOR HER DISCOVERY. THE FIBER IS PERHAPS BEST KNOWN FOR ITS USE IN BULLET- AND KNIFE-RESISTANT BODY ARMOR (FOR DOGS AS WELL AS HUMANS). MORE THAN 2,500 LIVES HAVE BEEN SAVED BY THE ARMOR. KWOLEK IS SHOWN HERE WEARING GLOVES OF KEVLAR®, WHICH ARE WIDELY USED TO PREVENT CUTS IN MANUFACTURING OPERATIONS. HUGE TIRES ARE REINFORCED WITH KEVLAR®, AS ARE SAILS. ROPES OF KEVLAR® ARE STANDARD EQUIPMENT ABOARD MANY OF THE WORLD'S SHIPS.

beginning of a new era for DuPont and other U.S. companies. New mores and regulations emerging from the civil rights, environmental and consumer movements soon governed every aspect of business life, from personnel hiring and promotion to manufacturing, marketing and waste disposal.

All things considered, the first waves of change washed fairly easily over the decks at DuPont in 1968. Net earnings rose for the first time in five years, although they remained lower than in any year since 1957. Prices held even after several years of decline, and wages and salaries passed the $1 billion mark for the first time in the company's history. DuPont's international businesses continued to prosper, confirming the general wisdom of that strategy. Exports rose 15 percent over the 1967 level, while sales of goods manufactured abroad exceeded the previous year's total by 19 percent. Construction began on a 19-story office building, DuPont's third headquarters facility in downtown Wilmington. It seemed like business as usual, including the successful introduction of a high-end, silky fabric called Qiana® nylon at haute couture fashion shows in Paris and New York. McCoy spruced up the company's staid Annual Report, adding photos, colors and new sections, but reminded stockholders that DuPont was in for more than cosmetic changes. "We live in troubled times," he wrote, "and public problems invariably have private consequences. Private industry cannot remain aloof."[15]

The environmental movement confronted DuPont and other chemical companies with their most sustained, difficult and costly challenge to traditional notions of corporate citizenship. The debate over industrial pollution was already old when the burgeoning chemical industry of the 1940s and 1950s created wholly new and often unexpected dangers of contamination. In 1962 biologist Rachel Carson's book, *Silent Spring*, which described the harmful effects of the insecticide DDT on fish and birds, raised new questions about a chemical that for nearly 20 years had meant protection from lice, mosquitoes and other disease-carrying pests. DuPont had

NOTABLE

> In 1962, DuPont was among the first major companies to sign the "Plans for Progress" statement, a voluntary agreement which vowed to not discriminate in employment on the basis of race, creed, color or national origin.

> The National Aeronautics and Space Administration's Mercury, Gemini and Apollo programs used DuPont products extensively. President Eisenhower had launched NASA in 1958, after the Soviet Union successfully orbited the satellite *Sputnik*. Three years later, President Kennedy set NASA the ambitious goal of putting a man on the moon "before this decade is out." On Sunday, July 20, 1969, Neil Armstrong stepped onto the surface of the moon's Sea of Tranquility. DuPont was proud of its many contributions to Armstrong's and NASA's triumph. The flag that the astronauts planted on the moon's surface was made of nylon. And DuPont materials such as nylon, neoprene, Lycra®, Nomex®, Mylar®, Dacron®, Kapton® and Teflon® constituted 20 of the 21 layers of the $300,000 space suits worn by Armstrong and his two fellow explorers, Edwin Aldrin and Michael Collins.

> Congress passed a huge tax cut in 1964, sure that a rising tide that lifts all boats would stimulate economic growth. At the same time, DuPont engineers and construction crews wrestled with rising tides of a different sort at low-lying sites such as the company's Dacron® plant at Cape Fear, North Carolina. Sometimes a dry sense of humor helped. One day early in the Cape Fear project the construction manager wired some good news to the Textile Department: "Congratulations! We have struck land on your property." DuPont engineers dealt with flooded construction sites overseas, too, but there they sometimes had to learn completely new construction techniques. Two-thirds of the company's planned Orlon® and Lycra® facilities at Dordrecht in the Netherlands lay under 14 feet of water. After the land was reclaimed by traditional Dutch techniques of diking and filling, long concrete piles with a large bulb on one end were driven deep below the surface in a process called "empoldering." Sometimes the 40-foot piles disappeared altogether in the subterranean ooze, but enough found a foothold to form a floating foundation for the entire plant.

> DuPont Far East opened sales offices in Singapore, the Philippines and Malaysia in 1973. Expansion continued in the Asia-Pacific region in 1976 with the formation of a new subsidiary to produce agricultural chemicals in Taiwan and the opening of the first representative office in Korea. And in 1981, DuPont opened a representative office for the Asia-Pacific region in New Delhi, India.

> DuPont established the first-ever accident response network in 1967 to expedite emergency response times for road, rail or sea accidents involving potentially hazardous chemicals. In 1971, DuPont helped organize an industry-wide transportation accident reporting and response system to give authorities faster reaction times for accidents involving hazardous materials.

> With the invention of the Permasep® hollow fiber reverse osmosis process in 1969, DuPont scientists were able to purify water to an extreme degree for industrial and medical applications. Subsequent improvements to Permasep® allowed for seawater desalinization, which garnered the company the coveted Kirkpatrick Chemical Engineering Achievement Award in 1971.

> Chemist David H. Woodward of DuPont's Photo Products Department initiated research in 1961 into solid photopolymer compounds — developed in the 1950s by Louis Plambeck and associates — for the manufacture of sophisticated printed circuit boards. The resulting Riston® line of dry photo resists dramatically cut circuit board chemical preparation times and became an immediate commercial success in the burgeoning electronics industry.

DuPont's Land Legacy Program releases large tracts of company-owned land to states for use as wildlife and natural vegetation preserves. Since 1983 DuPont has donated 50,000 acres — worth nearly $70 million — in the United States. They include 7,700 acres near Brevard, North Carolina, given in 1996 and known thereafter as the DuPont State Forest, and 3,300 acres on the Chesapeake Bay in 1997. Worldwide donations approach 300 square miles, including a large tract of land in the rolling hills of DuPont's Asturias, Spain, property. The company's Victoria, Texas, plant began a wildlife management program on 3,500 acres near the Guadalupe River in 1990, and between 1994 and 1998 DuPont created a wetland preserve on the property. The 53-acre wetland, including an education center, opened in August 1998 and soon became home to a proliferation of wildlife species, including bald eagles, great horned owls and whitetail deer. The Victoria Wetland won the National Wildlife Habitat Council's Corporate Habitat of the Year award in 1998 — the second such award for the Victoria plant's wildlife program in three years, and the first time a site had won the award twice.

been only one of many companies making DDT for the armed services during World War II. The company never made a profit on the product, however, and discontinued production in 1954. Still, Carson's call for environmental protection presaged a rising chorus of criticism of business-as-usual in America and helped speed passage of several antipollution and wilderness preservation laws in the 1960s.

For years DuPont had been aware of its environmental impact. In 1946, for example, the company had formed an Air and Water Resources Committee to address pollution problems resulting from its manufacturing operations. But environmental stewardship was a rapidly evolving notion in the 1960s, when many Americans sharply questioned the credibility of authorities generally and scientific experts specifically.[16] In 1960 only 124,000 Americans belonged to environmental organizations; by 1972 such groups claimed nearly 10 times that number. Public pressure mounted for legislation that would impose new standards on businesses and require them to help pay for what *U.S. News & World Report* estimated in 1970 would be a five-year environmental cleanup bill of $71 billion.[17] DuPont's share of that cost was estimated to be $600 million. On April 22, 1970, just three months after President Richard Nixon signed legislation establishing the Environmental Protection Agency (EPA), 10 million schoolchildren helped celebrate the first Earth Day. Politicians took notice. Between January 1, 1970, and the autumn of 1971, 3,000 environmental bills were introduced in Congress — 20 percent of all the new laws proposed during those months.

"We are in a new ball game," said McCoy in a 1972 speech to the Economic Club of Detroit. To succeed, however, both business and government would have to move beyond the adversary relationship implied by new regulatory agencies such as the EPA, the Occupational Safety and Health Administration, and the Consumer Product Safety Commission. Manager Samuel W. Fader of the Edge Moor pigments plant told a group of Delaware educators that running the plant was more like walking a tightrope than practicing precise science. The total of $6.5 million he had thus far invested in pollution abatement was one-sixth the book value of the entire facility. Fader struggled to meet environmental goals while keeping costs down, with an annual inflation rate of 4.5 percent tugging on both sides of his efforts.

In 1971 the United States recorded its first unfavorable balance of international trade since 1892. The new economic powerhouses of West Germany and Japan owed much to U.S. reconstruction initiatives after World War II. Now the Nixon administration moved to restore balance by jettisoning the provisions of the Bretton Woods agreements that had linked the value of other international currencies to the benchmark U.S. dollar. With the dollar off the gold standard, Nixon imposed nationwide wage and price controls in August, freezing DuPont's costs while allowing the company a 2 percent price increase. In 1972 DuPont's net earnings rose for the first time since 1968 as McCoy and the Executive Committee struggled to achieve the best mix in the company's formula for successful diversification. Acquisitions, internal restructuring and a new emphasis on entrepreneurship were key elements in their endeavors to adapt DuPont's size and resources to the rapidly changing conditions of a global economy.

DuPont had considered acquiring a pharmaceuticals firm as part of its long-term diversification strategy as far back as 1916, but its preoccupation with dyes in World War I had corralled most of the company's resources. In the 1930s Fin Sparre pushed DuPont to purchase Abbott Laboratories, but the Executive Committee settled instead on an agreement to have Abbott screen DuPont's products for any possible pharmaceutical benefit. But because DuPont's industrial departments were reluctant to turn over any new materials to an outside company, nothing useful came from the arrangement. During World War II DuPont challenged ICI's investment in penicillin research as a violation of the two

UNITED STATES
ENVIRONMENTAL PROTECTION AGENCY

companies' patent-sharing agreement but soon reconsidered the consequences of a possible breach with the British firm. Heeding Charles Stine's counsel, DuPont withdrew its protest and moved instead into veterinary pharmaceuticals, hoping that such work might ultimately yield benefits for humans.

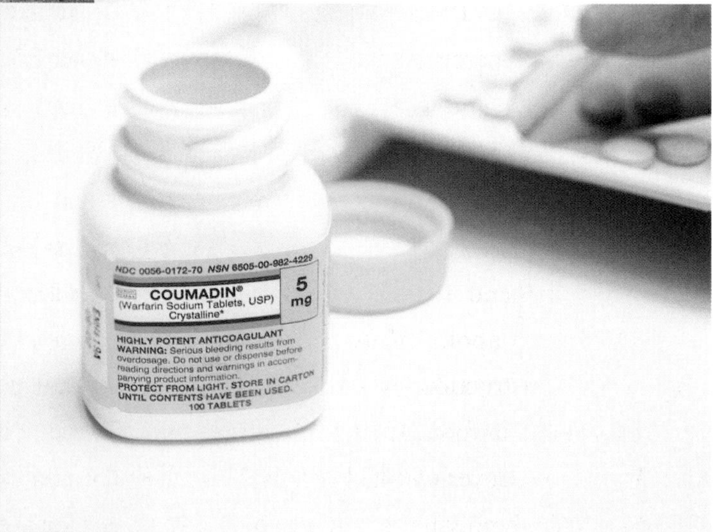

Finally, in December 1969, at the urging of Edward R. Kane, a chemist and former general manager of the Industrial and Biochemicals Department who had joined the Executive Committee earlier in the year, DuPont purchased a New York pharmaceutical company, Endo Laboratories. It was DuPont's first acquisition in nearly 25 years. DuPont had stopped active promotion of its influenza treatment drug, Symmetrel®, early in 1969. The drug had not sold well, partly because the market for preventives was stronger than that for treatments and partly because DuPont's unfamiliarity with the arcanum of FDA approval had cost it valuable time in making Symmetrel® available for the flu season. Kane and others believed that acquiring Endo Labs, makers of the successful anticoagulant Coumadin®, would help DuPont overcome the marketing and regulatory problems that its efforts with Symmetrel® had highlighted. In the next two years, DuPont extended its cautious acquisitions strategy beyond pharmaceuticals into other promising new fields. The company purchased Berg Electronics and bought Bell & Howell's analytical instruments line, which featured mass spectrometers as well as moisture and leak detectors. The company passed a historic milestone in 1971 when Polymer Intermediates finally discontinued commercial manufacture of DuPont's original product, black powder, and of dynamite. Tovex® water gel explosives were the last remnant of the company's product heritage.

McCoy established "profit centers" in 1968 to encourage entrepreneurial initiative and bottom-line accountability among personnel directly involved in a variety of targeted product lines. He also brought the empyrean image of the Executive Committee closer to earth in 1971 by assigning each member direct monitoring responsibility for one of the 12 industrial departments. McCoy's streamlining reduced the company's overall workforce by 10 percent in 1970-1971, mostly through retirements and attrition. McCoy also oversaw DuPont's merger with Christiana Securities, an act that dissolved the last trace of organized, du Pont family control over DuPont. Legal challenges held up final Securities and Exchange Commission approval of the deal until 1977, when DuPont exchanged its stock for that held by Christiana Securities.

At first glance, 1973 seemed to signal a phenomenal recovery for DuPont. The company's sales topped $5 billion for the first time, and net earnings also scored a record total of $586 million, 41 percent higher than 1972's level. About half of this profit came from the company's international businesses, which also achieved a record of $1.6 billion in sales. Inflation swelled the numbers, and the devalued U.S. dollar contributed to the international sales increase, as did worldwide demand for almost all of DuPont's products. Despite alarms about rising energy costs, the company announced construction of a new, automated dye plant in Puerto Rico and introduced an extraordinary new fiber called Kevlar®. The discoveries of Stephanie Kwolek in the area of liquid crystalline polymer solutions formed the basis for the commercial preparation of Kevlar®. Fifteen years and $500 million in the making, Kevlar® had five times the tensile strength of steel, but the fiber's high production costs delayed its ultimate market success for several more years.

DuPont's sales surge in 1973 also reflected anxiety among its customers, who stockpiled all kinds of products to hedge against imminent price hikes and rumored energy shortages. McCoy told stockholders that the company's success that year had "possibly borrowed from the future."[18] It had. But so had the United States. In the early 1970s DuPont, like America, had grown more dependent on overseas sources for its energy and raw materials needs. The nation's daily oil imports nearly doubled in just three years, from 3.2 million barrels in 1970 to 6.2 million barrels in 1973. Escalating tensions between Israel and its neighboring Arab states erupted into war. On October 6, 1973, Egyptian and Syrian forces invaded Israel on the Jewish holy day Yom Kippur. When the Israelis successfully repelled the invasion using U.S. and Allied military supplies, the Organization of Petroleum Exporting Countries (OPEC) imposed a retaliatory embargo on Israel's supporters. In December OPEC stiffened its embargo by raising oil prices to $11.65 a barrel, nearly a 400 percent increase in just three months.

Never before had DuPont been so severely pressured to make hard choices about its present health and ultimate survival. Never before had those choices depended so much on managing political and social factors beyond the company's direct control. As a succession of blows threatened to enervate the national economy, DuPont chose a new leader who had no professional business or scientific background, but who impressed many in the company with a special wisdom that the times now demanded. DuPont had always prided itself on its willingness to take great risks. No one now appreciated those risks better than Irving S. Shapiro, who succeeded Charles McCoy as the company's chief executive officer on January 1, 1974.

The 1969 purchase of Endo Laboratories brought to DuPont a pharmaceutical line that included anticoagulants, antitussives and antispasmodics. Narcan® naloxone hydrochloride for reversing undesirable side effects of narcotic analgesics was introduced in 1971. A nasal decongestant and a mild sedative were introduced in 1972. Use of Symmetrel® amantadine hydrochloride for Parkinson's disease was approved in 1973 by the Food and Drug Administration.

CHAPTER 8

A SEARCH FOR DIRECTION

INSET RIGHT: **Irving S. Shapiro (1916–2001) became chairman and chief executive officer in December 1973. He restructured the firm's long-term debt to fund new plant capacity and also reoriented DuPont into specialized, high-return products such as agricultural chemicals. He maintained the company's core commitment to research while shifting its research and development focus toward product lines not dependent on volatile petroleum supplies.**

BELOW: **Forty years of DuPont leadership — 1940 to 1980 (clockwise from lower left): Crawford H. Greenewalt, Lammot du Pont Copeland, Irving S. Shapiro, Charles B. McCoy and Walter S. Carpenter Jr.**

"The race is not to the swift nor the battle to the strong," says the preacher in Ecclesiastes (9:11), "but time and chance happeneth to them all." Time and chance, and more, happened to DuPont during the energy crisis triggered by the Organization of Petroleum Exporting Countries (OPEC) oil embargo in the winter of 1973–74. In the months before his retirement, and before the embargo, Charles McCoy had developed, with the Executive Committee, a plan for getting the company through the uncertain times ahead. The appointment of attorney Irving S. Shapiro on January 1 to DuPont's top executive position as chairman and chief executive officer was an explicit recognition of changing times. "What people want to know," said Shapiro, "is not what the company is doing commercially, but simply what kind of people are running the show — Who are you and what do you believe in?"[1]

Shapiro had learned early in life that the rule of law and a free press were more than fine ideals. They were effective ways to fight for change. In the 1920s Shapiro's father had been pistol-whipped in his Minneapolis dry-cleaning store for refusing to participate in a citywide price fixing racket. The corruption reached into city hall, the police department and even city newspapers, with one exception — the *Saturday Press*, whose publisher saw in the Shapiro case an opportunity to scorch his many enemies in high places. When the county attorney tried to suppress publication, the American Civil Liberties Union and the *Chicago Tribune* promoted the cause as a First Amendment issue. Shapiro's side eventually won before the U.S. Supreme Court in *Near v. Minnesota* in 1931.[2] Twenty years later Irving Shapiro himself appeared before the Supreme Court as a young Justice Department lawyer in *Dennis v. United States*, a First Amendment case involving violations of the anti-communist Smith Act. Shapiro's arguments carried the day for the government and captured the attention of one of DuPont's lawyers, Oscar Provost, a former Justice Department attorney, who soon recruited Shapiro for the company's Legal Department.

Shapiro guided DuPont through the tribulations of the General Motors antitrust suit in the 1950s and early 1960s. As a member of the Executive Committee in the early 1970s, he upheld the free speech principles learned in his boyhood by shepherding a group of eight "Raiders" from Ralph Nader's advocacy group through interviews with company executives. Nader's Center for Study of Responsive Law chose DuPont because the company's reputation as a progressive firm suggested it would welcome consumer advocates, and because its political and economic influences would be more evident in its small home state of Delaware. Indeed, McCoy and Shapiro were accommodating. But they were not naive. Shapiro accompanied the young investigators — "sometimes in error but never in doubt," he later said of them — through all their interviews, offering his counsel when necessary. Unfortunately, the Nader group's conclusions were

LEFT: **The company was led in 1975 by Chairman Irving S. Shapiro (seated center) and an Executive Committee consisting of Edward R. Kane (seated left), Irénée du Pont Jr. (seated right), and (standing left to right) Norman A. Copeland, Joseph A. Dallas, Edwin A. Gee, Richard E. Heckert, Edward G. Jefferson and Samuel A. Milliner Jr., secretary to the committee.**

CENTER: **Theodore L. Cairns became director of central research in 1971 at a time when many in management wanted more emphasis on the "D" in "R&D." The appointment of Cairns signaled that DuPont still placed great emphasis on purely scientific research.**

RIGHT: **Edward R. Kane became president and chief operating officer in 1973. The energy crisis and increasing global competition presented the greatest challenges to Kane and Chairman Irving S. Shapiro. They responded by refocusing DuPont on non-petroleum related products like electronics, agricultural chemicals and pharmaceuticals.**

widely criticized as foregone and biased, and the study had little positive effect on DuPont or Delaware.[3]

In other areas Shapiro's diplomacy proved more successful. He firmly but delicately pushed Lammot Copeland and the Executive Committee to bring the disappointing Corfam® poromeric venture to a close, and he earned additional respect for his judicious balance of DuPont company and du Pont family interests during the Christiana Securities merger. The forward-looking McCoy did not foresee the OPEC embargo of 1973–74, but he knew that DuPont would need the new perspective that Shapiro brought to the company's top leadership. McCoy also knew that Shapiro himself would use wisely the operational and management experience of a chief operating officer (COO). In July, several months before Shapiro became CEO, McCoy named chemist and former general manager of the Industrial and Biochemicals Department Edward R. Kane to fill a new office — president and COO of DuPont.

By 1974, 70 percent of DuPont's products were made from petroleum-based raw materials. That stark chemical fact of life posed a monumental challenge for DuPont executives. As the company's demand for energy and for raw materials rose, so did federal regulatory pressure for conservation and cleaner production. The goals of regulation were consistent with DuPont's own goals in terms of both cost savings and pollution abatement, but the company often found itself at odds with federal authorities over the best means of achieving them. At the same time, rising energy costs in the 1970s compounded the double-digit inflation ignited by the Vietnam War's $150 billion price tag. In 1974 and 1975 DuPont's faltering finances dwarfed other issues competing for company executives' attention.

DuPont's net income fell 31 percent in 1974 as its manufacturing operations absorbed almost a billion-dollar increase in energy and raw materials costs — a devastating rise of 80 percent in just one year. Nationally, inflation neared 12 percent in 1974, and resulting slumps in the housing construction and automobile markets cut deeply into DuPont's profits. The company's record-high sales figures, up 16 percent from 1973, offered only illusory comfort, for they came not from increased volume but from higher prices following the government's final lifting of wage and price controls in April after those measures had failed to stem inflation.

DuPont plant managers reduced their energy consumption by 4 percent in 1974, adding to the 30 percent they had saved between 1968 and 1973 in response to company-wide directives to lower energy costs. Now Shapiro and the Executive Committee sought additional savings and efficiency by combining the Central Research Department and the Development Department into one, Central Research & Development (CR&D), under director Ted Cairns. They also slowed international expansion while looking for ways to "integrate backwards" into oil production in order to gain some control over the company's raw materials or "feedstock" supplies. DuPont's 40 percent investment in 1974 in the Polyacryl Iran Corporation, a polyester and acrylic fibers maker in Isfahan, Iran, was the company's first venture in the Middle East. It presented an opportunity to further the company's long-term strategy of internationalization while reducing raw materials costs.

In 1975 the company's net earnings fell another 33 percent during the nation's worst economic recession since the Great Depression. Even excluding the effects of inflation, DuPont's 1975 earnings were its lowest since 1953. Its overall rate of return on investment dropped from 7 percent in 1973 to 2.5 percent in 1975 as worldwide industry overcapacity, competition and soaring raw materials costs flattened DuPont's once booming fibers business. But the explosives business enjoyed its best year ever as oil, coal and gas companies expanded their drilling and mining operations to meet demands for increased domestic production.

While construction crews snaked 800 miles of oil

DuPont explosives — such as Pourvex® water gel — were used extensively in the construction of the Trans Alaska pipeline. The 48-inch pipeline has 11 pumping stations along its 800 miles that move about 47,000 gallons of crude oil per month at a velocity of 6 mph.

BELOW: Irving S. Shapiro (right) and Joseph Wyke worked together with United Way to aid the Wilmington, Delaware, community in the 1970s.

pipeline across Alaska, from Prudhoe Bay on the North Shore to the port of Valdez on Prince William Sound, Shapiro searched for cheaper oil and gas feedstocks. The pipeline tapped into an estimated reserve of 10 billion barrels, but the oil would not reach Valdez until 1977 and would not meet all of DuPont's needs. The company explored possible agreements with Atlantic Richfield and Shenandoah Oil and entered a joint venture with National Distillers and Chemical Corporation to produce raw materials in Deer Park, Texas, for DuPont's methanol plants. At the same time, Shapiro sought to increase DuPont's maneuvering room by opposing "excessive regulation" by the federal government, by seeking extension of federal timetables for meeting clean air and water standards, and by challenging the scientific bases for theories about depletion of the ozone layer as well as for the U.S. Environmental Protection Agency (EPA) ban on lead additives.

Shapiro's reputation, however, rested mainly on his ability to forge positive solutions from adversarial interests. In the early 1970s he joined several *Fortune* 500 CEOs in the newly formed Business Roundtable, seeking a middle ground for executives pulled between a "take no prisoners" hostility toward taxes and regulations on one side and resigned capitulation to governmental authority on the other. As Shapiro saw it, the Business Roundtable fought for representation in the councils of power and policy-making in Washington after years of what had seemed like mutual hostility between business and the federal government. He rejected the idea that government was basically a business run badly, but he also denied that business should become just another interest group "tugging at the sleeve of Congress, pleading for favors."[4] He argued that government and corporations were both so large and so influential that they needed to rise above short-term interests, take a longer view and work together to make sensible policies.

In 1977 DuPont's Legal Department started a full-time lobbying effort, the Governmental Affairs Action Program, to inform legislators and officials about the company's position on such matters as environmental and industry regulation. The following year the company's general counsel, Charles E. Welch, who acknowledged there had been a time in his life when he "didn't know ambient air from a billy goat," filled a new office, Vice President for External Affairs, to handle DuPont's increased lobbying, legal and public relations activities. The Business Roundtable did not invite organized labor to its discussions. Business and labor remained deeply divided on issues like outsourcing, downsizing, health insurance and automation. Nevertheless, DuPont prided itself on a history that was remarkably free of labor troubles, and Shapiro sometimes joined labor leaders like the AFL-CIO's George Meany and Lane Kirkland in a different venue, the Labor-Management Group, to discuss issues made more contentious than ever by the weakened national economy and shrinking energy resources.[5]

Shapiro also kept DuPont in pace with changing social attitudes, thereby honoring in new ways the company's long tradition of community responsibility. His nominations of retired Federal Reserve Board governor Andrew Brimmer and biologist Ruth Patrick for election to DuPont's Board of Directors in the mid-1970s led to their selection as the board's first African American and first woman members. By 1979 the company maintained 250 affirmative action programs nationwide and deposited funds in 24 minority-owned banks. About 30 percent of the college graduates DuPont hired in 1979 were minorities or women. At the local level, DuPont

The Committee on Audit of the Board of Directors in 1979 reflected the diversification of the company that intensified throughout the 1970s. Committee members were (seated) Ruth Patrick and Charles L. Brown and (standing left to right) William Winder Laird, Caryl P. Haskins, Gilbert E. Jones and Andrew F. Brimmer.

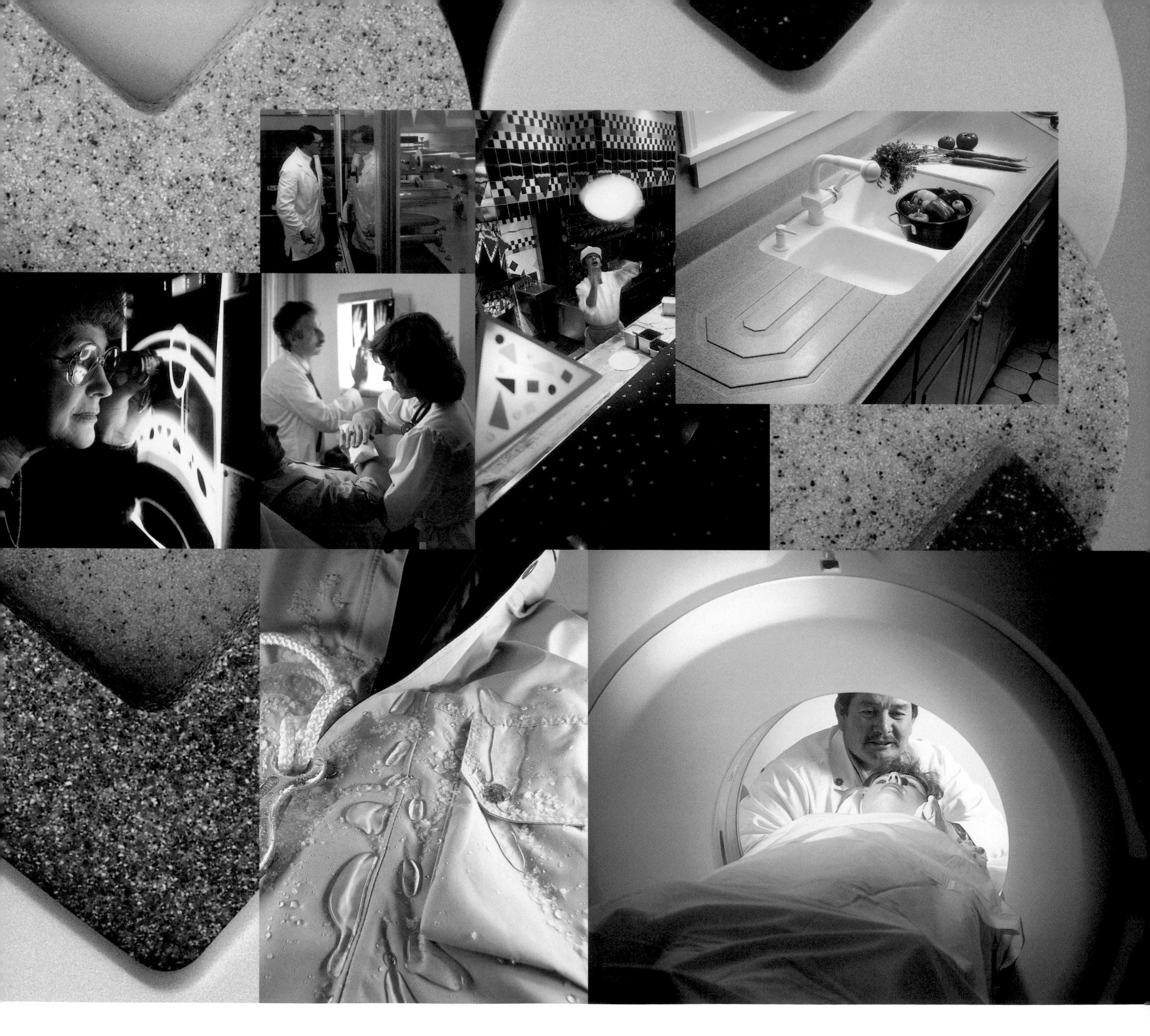

participated in Wilmington's Forum for Advancement of Minorities in Engineering, and Shapiro often met personally with Wilmington school officials to help them find the funds and supplies, or sometimes just the public support, they needed. More than 50 DuPont attorneys quietly volunteered their services as public defenders in the Wilmington courts. But DuPont was still above all a for-profit enterprise, and 1975 had been a bad year for business.

Both Shapiro and Kane publicly hailed the nation's bicentennial year, 1976. "It will be a far better year," they reassured stockholders. They were right. Over the next four years DuPont's net income rose steadily to nearly $940 million, a cumulative gain of 150 percent over 1975's low of $272 million. The company's nonfibers products, now called Chemicals, Plastics and Specialty (CPS) products, performed well, as did the Automatic Clinical Analyzer medical diagnostic system. Electronics and crop protection products also contributed to DuPont's improved earnings, while two-thirds of the company's 80 New Venture products of the 1960s, such as Corian® surface materials, finally reached profitability. In spite of economic hard times, American consumers maintained a healthy appetite for many of DuPont's products.

DuPont's international operations showed impressive gains in the late 1970s, despite Iran's takeover of Polyacryl Iran Corporation during that country's fundamentalist revolution in 1978. By the end of 1979 DuPont's overall return on investment had bounced back to 6.2 percent from its 1975 low of 2.5 percent. Even fibers products had rebounded to account for 31 percent of DuPont's net earnings in 1979, compared with only 2 percent in 1975. But this was still far below fibers' net 1971 earnings, which had accounted for nearly half of the company's total.

DuPont's executives and employees had wrung the profits of 1976–79 from the company's own resources as well as from the marketplace. Shapiro, Kane and the Executive Committee pared R&D professional staff by 20 percent

OPPOSITE: (clockwise from top left) A coater in Neu Isenburg, Germany, made experimental versions of Riston® photoresists.

Corian® surfaces have evolved from pale countertops to eye catching design elements in commercial and residential settings.

DuPont purchased New England Nuclear Corporation (NEN) in 1981 as part of its expanding venture into medical research and health care. NEN radiopharmaceuticals were critical to new imaging modalities.

Teflon® stain protection defends fabrics against water and oil-based stains.

DuPont nondestructive testing x-ray film ensured product safety. The company pioneered numerous medical x-ray products, including mammography film for the detection of early breast cancer with a low radiation dosage. DuPont left the x-ray film business in 1997.

LEFT: Corian® was introduced in 1967 in three colors. Today, the palette numbers more than 90.

DuPont first turned its attention to agriculture in the early 1900s with the investigation of nitrogen fixation by plants, and in the 1920s the company produced seed disinfectants on a small scale. But it was DuPont's purchase of the Grasselli Chemical Company and the R&H Chemical Company in 1928 and 1930, respectively, that established its expertise in inorganic insecticides and fungicides. Agricultural research during World War II spurred a "chemical revolution" in the postwar decades, when DuPont introduced new synthetic, organic herbicides such as Telvar® and Karmex®, and insecticides such as Lannate®.

between 1974 and 1977 and reduced funding for basic research from one-third to one-quarter of the total R&D budget. Overall research funding increased during these years — though not as much as inflation — but more of it now went to improving products and manufacturing processes rather than to open-ended discovery projects. Chemist Howard E. Simmons, who succeeded Ted Cairns as director of CR&D in 1979, recalled how Irénée du Pont Jr., senior vice president and great-great-grandson of E.I. du Pont, tried to boost sagging morale among Experimental Station scientists during the "contraction years" of the 1970s. "He'd be waiting for me at eight o'clock in the morning, and the gist of this was to say, 'Hang in there. Things are going to get better. Don't worry about the way things are going.'" Simmons did not believe that Shapiro was going to cut research drastically, but, he thought, "people were just still not sure where in the hell they were going."[6]

Researchers like George Levitt plowed their special fields of interest despite uncertainties about the future. Sometimes they turned up new findings that underscored DuPont's historic reliance on the ultimate profitability of basic scientific knowledge. In 1975, following DuPont chemist Conrad Hoffman's work on plant enzymes in the 1950s, Levitt discovered a remarkable group of chemicals called sulfonylureas that interfered with an enzyme vital to plant growth. Very small amounts of sulfonylureas — "thimbles, not barrels," DuPont boasted — proved to be potent herbicides. They were also relatively safe, having no effect on rain runoff and causing no contamination of groundwater. And since the enzyme they blocked existed only in plants, the sulfonylureas posed virtually no danger to mammals. DuPont patented Levitt's discovery in 1978 and four years later introduced its Glean® herbicide to wheat farmers. Before long the company marketed sulfonylurea herbicides for every major food crop in the world.

In 1978 the Executive Committee assigned one of its members, Edward G. Jefferson, to take direct responsibility for

DU PONT
DU PONT

Turbo
WARNING

the company's worldwide research activities. Shapiro and Jefferson, a University of London chemist who had been recruited by DuPont's old Polychemicals Department in 1951, struck a positive note by reaffirming one of DuPont's traditional strengths. "Our company lives by high technology," they said in 1979 while they streamlined several of the company's operations. In 1978 two departments were combined to form the new Chemicals, Dyes and Pigments Department, and a new Petrochemicals Department absorbed the company's beleaguered Freon® business. The next year the Plastic Products and Resins Department merged with Elastomer Chemicals to form a new Polymer Products Department. The Spruance cellophane plant near Richmond, Virginia, ceased production in 1976, and in mid-1979 DuPont announced that it would stop producing dyes, a particular disappointment to workers at the company's four-year-old, state-of-the-art dye plant in Manati, Puerto Rico. But these cutbacks were not enough to counter the stubborn, double-digit inflation of the 1970s; for the first time in its history, DuPont had to borrow money to finance capital growth and improvements.

After more than a year as the Executive Committee's research monitor, Edward Jefferson succeeded Edward Kane in January 1980 as company president and chief operating officer, while physicist William G. Simeral took on Jefferson's former duties on the Executive Committee. Kane's seven years as president had capped a highly successful, 36-year career at DuPont, though he had probably been as surprised as others in the company when McCoy and the board named Shapiro CEO in 1974. Now Jefferson, not Kane, seemed favored to become DuPont's next chief, and Kane chose to retire.

During 1979 Jefferson had closely studied the company's energy needs. It seemed clear to him that DuPont would be wise to continue its earlier effort to "integrate backwards," that is, to own its own petroleum resources rather than purchase them on the volatile open market. DuPont's prior attempts at shared energy ventures, with Atlantic Richfield and Shenandoah Oil, had not materialized, but in 1979 DuPont signed a five-year agreement with the Continental Oil Company (Conoco) to explore for natural gas in Texas. In 1980 the two companies broadened the agreement to include oil and gas searches in Louisiana and Mississippi.

Meanwhile, the Iranian revolution had created new troubles in the Middle East. The OPEC cartel raised oil prices, increasing DuPont's costs by 36 percent in 1980 and exacerbating the new economic doldrums, dubbed "stagflation" — stagnant growth, rising prices — afflicting the American economy. Political forces affecting DuPont's businesses seemed beyond the company's control, but Shapiro's determined diplomacy often shaped unfolding events to the company's advantage. In 1977 Shapiro, DuPont's first Jewish CEO, worked with Congress on legislation that helped defuse a threatened Arab boycott of companies doing business with Israel. Then he traveled to Saudi Arabia and Israel with businessman George Shultz — later to be secretary of state — to explain the positions of the U.S. government and U.S. business on the boycott issue.[7]

But the most enduring achievement of Shapiro's cooperative approach to policy making came in December 1980, President Carter's last month in office as well as Shapiro's last month as DuPont's CEO. Faced with polls showing overwhelming public support for a law that would mandate cleaning up of toxic waste sites, members of the Chemical Manufacturers' Association (CMA) — now called the American Chemical Council — had struggled for months over the best way to respond to the mounting pressure for change. DuPont's William Simeral, who was also on the association's Board of Directors, recalled one CMA meeting in which a participant opposed to cleanup legislation proposed an advertising campaign emphasizing the industry's positive contributions to society. A colleague replied, "Well, that's all well and good, but don't you think it would be a good idea if we had some results before we started such a campaign?"

DuPont herbicides like Glean® and Accent® belong to a family of crop protection chemicals known as sulfonylureas, which kill weeds by inhibiting a plant enzyme necessary for their growth. Since only plants have this enzyme, the sulfonylureas are relatively safe for animals and humans. Sulfonylurea herbicides act directly on plant surfaces rather than through absorption by roots. And since only small amounts are required to do the job, runoff and seepage into groundwater are greatly reduced when compared with older products. DuPont chemist Conrad Hoffman discovered the basic mechanism of ureas on plant enzymes in the 1950s, and another DuPont researcher, George Levitt (far left), discovered the potent sulfonylureas in 1975. DuPont patented the substances as herbicides in 1978 and first marketed one of them, Glean®, to wheat farmers in 1982. Sulfonylureas have since been developed for every major food crop in the world. Levitt retired from DuPont in 1986 after 30 years of service. He earned DuPont's Lavoisier Medal and in 1994 was awarded a National Medal of Technology by President Bill Clinton.

"That shut everybody up," Simeral recalled with a chuckle.[8] For months DuPont and other companies had worked with Congress to fashion a workable cleanup bill. Now, as the final vote drew near, unity and cooperation within the industry were crucial to the bill's success. If the effort unraveled, industry supporters of toxic waste cleanup would have to start from scratch, with their public image damaged even further by the delay. When Shapiro took the lead in saying so to a reporter from the *New York Times*, he was joined by representatives from several other chemical companies, and Congress was at last emboldened to pass a bill.[9]

In December 1980 President Carter signed the Comprehensive Environmental Response, Compensation and Liability Act, often referred to as CERCLA, or the Superfund cleanup bill. CERCLA required a number of companies, including DuPont, to provide $1.6 billion over five years as a trust fund for cleaning up toxic waste sites. Looking back on DuPont's role in achieving unanimous support for the bill among companies that often had very different points of view, Simeral recalled, "I felt pretty good about that."[10] So did President Carter, who singled out Shapiro for special praise at a White House ceremony for his contributions to the bill's timely passage.[11]

CERCLA may well have served DuPont's long-term interests, but in the short run it did little for the company's bottom line. The prices DuPont charged for its products rose about 7 percent each year in 1979–80, but its costs rose higher still as the annual inflation rate for the national economy swelled to a staggering 15 percent. President Carter appointed economist Paul A. Volcker to chair the Federal Reserve Board in 1979, aware that Volcker's proposed strong medicine — high interest rates — not only would stifle inflation but might dampen the overall economy as well. Indeed, Volcker and the Fed's Board of Governors did just that, raising interest rates so high that they triggered a sharp recession that lasted until 1983.[12] The national unemployment rate reached 11 percent in

the recession, the highest rate since the 1930s, and in 1982 DuPont plants cut back production to 65 percent of capacity.

DuPont's earnings sank, and eventually rose again, with the national distortions caused by the Fed's monetary policies. But DuPont's executives did not wait passively for macroeconomic solutions to emerge from Washington. In 1980 the company's fibers business cut its losses on commodity products by shutting down Orlon® acrylic fiber manufacture at DuPont's Maydown plant in Northern Ireland. At the same time, its $200 million expansion of Nomex® brand fiber and Kevlar® brand fiber production at its Spruance plant near Richmond, Virginia, was one of the largest capital appropriations in company history. DuPont also found new markets in Europe for its Riston® dry film resists and Kapton® polyamide film and for its Curzate® fungicide for controlling mildew on grapes. On the other side of the globe, a new electronics plant in Singapore enlarged the company's share of the fast-growing Far East and Asian markets, and in April 1981 DuPont acquired the New England Nuclear Corporation in an effort to strengthen its life sciences research capabilities.

The next month 65-year-old Irving Shapiro retired after seven years as DuPont's chairman and CEO, though he remained on the board as a member and chair of the Finance Committee. Shapiro's counsel was still available when time and chance presented his successor, Edward Jefferson, with one of the boldest and most difficult propositions the company had ever considered. Earlier that spring a small Canadian oil company, Dome Petroleum, had made an appeal directly to Continental Oil Company shareholders — a tender offer — to purchase a Conoco subsidiary. The surprisingly positive response Dome received from those shareholders signaled to larger investors that Conoco might be vulnerable to a hostile takeover. Executives from another Canadian firm, the Seagram Company Ltd., headed to Conoco's Stamford, Connecticut, headquarters to discuss a deal with Conoco's CEO, Ralph E. Bailey. But when air crews for Seagram's

DuPont purchased petroleum manufacturer Conoco, Inc. in 1981. The purchase gave DuPont a secure source of petroleum feedstocks needed for many of its fiber and plastics operations. Conoco also manufactured profitable commercial petroleum products and coal, produced by the wholly owned subsidiary Consolidated Coal Company. DuPont sold all of its Conoco shares in 1999.

LEFT: **Conoco made a major find of oil and gas in the Norwegian Sea in 1985. Named the Heidrun field, it began production in 1995 through a massive tension leg platform.**

Sorvall® centrifuges were a mainstay of medical research.

Virology was a major field of research for DuPont Pharmaceuticals. DuPont left the pharmaceutical business in 2001.

DuPont introduced Cyrel® flexographic printing plates in the early 1970s, primarily for the reproduction of packaging graphics.

Edgar Bronfman overheard frequent references to Wilmington from the airport control tower, they correctly inferred that Bailey was also meeting with DuPont.[13]

Over the summer several contenders besides Seagram entered bids for Conoco. But Bailey wanted to sell to DuPont, a company he already knew and trusted, rather than to a relative unknown such as Seagram, which had recently reaped handsome profits by selling the Texas Pacific Oil Company that it had owned for many years. A purchase by DuPont would serve both companies, offering Conoco the prospect of a reasonably predictable future while giving DuPont some protection from the severe economic cycles that volatile energy supplies could trigger. Arguments for the acquisition were compelling; the recent recession had reminded DuPont's leaders once again of the company's vulnerability to the vagaries of the international oil trade. But there also were grounds for serious reservations about a purchase so large that it was likely to change DuPont greatly, and perhaps not for the better.

Over several tension-filled weeks in August and September, Jefferson helped persuade DuPont's leaders to take a giant step toward ending the company's dependence on external oil supplies. DuPont researchers, he argued, could help Conoco find new ways to recover oil reserves, and Conoco's earnings would help stabilize DuPont's earnings through down-cycles in the global, energy-dependent economy. On September 30, DuPont and Conoco closed the deal, making Conoco — lock, stock and oil barrel — a wholly owned, separately operated DuPont subsidiary. Conoco, the world's ninth largest oil company, brought into the deal the valuable assets of its own subsidiary, Consolidated Coal, Inc. (Consol), thereby giving DuPont control over an alternate power source for its manufacturing plants. The $7.8 billion purchase, half in cash and half in stock, was the largest business deal in U.S. history up to that time and boosted DuPont's ranking from the 15th to the seventh largest U.S. industrial company.

One aspect of the agreement, however, had long-term

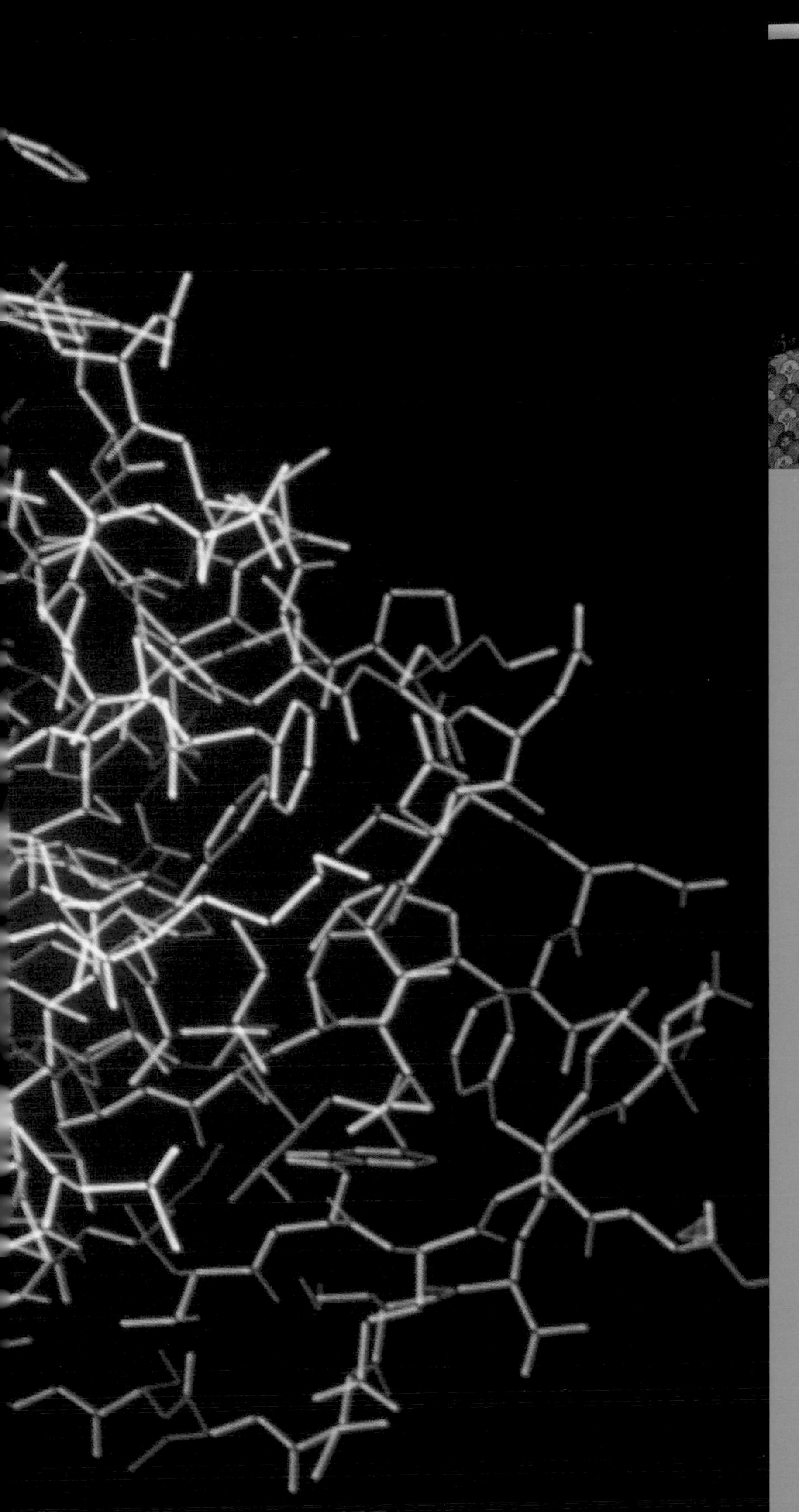

Edward G. Jefferson (left) served as chairman from 1981 to 1986. He was succeeded by Richard E. Heckert, who served until 1989.

implications for DuPont's management. The stock half of the purchase allowed Seagram to get in on the "back end" of the deal by exchanging its tendered Conoco shares for DuPont stock. The offer appealed to some investors, who saw opportunity in the expected drop in DuPont stock value after the deal as a result of the drag of new debt on the company's short-term growth and profitability. There was little DuPont could do about this development. Seagram's resulting acquisition of 24.5 percent of DuPont common stock gave Edgar Bronfman, his brother Charles and Seagram executive Harold Fieldsteel a kind of consolation prize for their rejection by Ralph Bailey; all three subsequently joined Bailey at the great oval table in DuPont's board room, where 27 other directors now presided over the affairs of a suddenly transformed firm.

To ease anxieties about Seagram's unanticipated presence at the table, Seagram allowed Jefferson and Shapiro to join its Board of Directors. The two companies also arranged a "standstill" agreement limiting Seagram's DuPont holdings to no more than 25 percent of DuPont's outstanding stock and giving DuPont first right of refusal if Seagram decided to sell its DuPont shares. But Jefferson found it necessary to keep explaining the purchase to skeptics, both inside and outside the company. There was no contradiction, he said, between chemistry and energy. DuPont's new COO, textile fibers veteran Richard Heckert, put it in an automotive metaphor. Conoco's oil and coal, he said, added "two more strong cylinders to go with the eight we had," the eight being DuPont's existing industry segments.[14]

DuPont had purchased Conoco in part to obtain relief from the on-again, off-again production restraints and price uncertainties of OPEC-controlled oil supplies. But OPEC itself soon suffered from cycles produced by its own price fixing strategy. Oil companies like DuPont's Conoco challenged the cartel by developing new oil and gas reserves. Automobile companies manufactured more fuel-efficient cars, and industrial consumers switched from oil to coal to

In the late 1970s DuPont sponsored a pedal-powered aircraft named *Gossamer Albatross* (BOTTOM PHOTO), a lightweight plane that promised to achieve the first human-powered flight over the English Channel. Dr. Paul MacCready's plane, made with Mylar® and Kevlar®, gave the company a chance to step back from the image of heavy industrial production and bureaucracy and soar on individual effort into clean skies. The *Gossamer Albatross* was made mostly of synthetic materials, almost all plastic save a few metal parts. Its sole pilot was Bryan Allen, who used only muscle power to pedal the propeller of the *Albatross*. After years of preparation and months of waiting through bad weather, the DuPont team joined the plane's engineers in witnessing a suspense-filled historic journey. On June 12, 1979, the fragile-looking *Albatross* lifted slowly off English soil and headed out over the waters bound for France. The trip kept everyone on edge as Allen touched the water several times, appeared to be exhausted, and even signaled for help at one point. However, after a breathtaking two hours and 49 minutes, Allen safely landed the plane.

fuel their power plants. DuPont shifted its Cape Fear, North Carolina, and Chambers Works, New Jersey, plants to coal in 1982. With half of the company's plants now using coal, DuPont saved the equivalent of 4.7 million barrels of oil per year. Such adaptations forced OPEC to lower oil prices in the 1980s, but this development called into question one of DuPont's main justifications for the Conoco deal. What was the point of a $7.8 billion oil company purchase if the higher costs it was intended to overcome never materialized?

Jefferson and Heckert continued to defend the Conoco purchase. They argued that agreements just like the one DuPont had made with Conoco greatly helped to pressure OPEC into lowering its prices. Besides, DuPont had to move in new directions, away from its old identity as simply a chemical company. Citing diminished growth in the chemical industry during the 1970s, Jefferson claimed that "DuPont no longer fits the traditional chemical industry definition. Our company is based on discovery."[15] "Discovery" had a nice rhetorical ring. It was positive, upbeat and consistent with DuPont's rich research history. But in the 1980s, a decade of corporate raiders, junk bonds, leveraged buyouts and hostile takeovers, it applied as much to DuPont's evolving identity as to the discovery of knowledge — or of oil.

Increasingly, DuPont shifted its R&D efforts away from oil-dependent products and into dramatically different fields. These included electronics and a diverse group of "life sciences" such as molecular biology, virology, pharmaceuticals and agriculture. There also was greater emphasis on lightweight, energy-saving polymers for automobile and aircraft construction. Though DuPont remained skeptical about many federal standards and regulations, the company's engineering plastics businesses profited from the demand for lighter, fuel-efficient cars. By 1980 the average American-made car contained 200 pounds of plastics. It was not very clear at the time, but DuPont's separate research efforts into polymers and molecular biology were headed toward eventual convergence, along the very lines

of "discovery" that Jefferson proposed.

In 1953, two years after DuPont recruited Jefferson, Britain's Francis Crick and the young American scientist James Watson published their discovery of the molecular structure of DNA, the protein building block of all living matter. Crick later wrote a book arguing that the aim of modern biology was to explain life processes in terms of physics and chemistry.[16] Jefferson was much impressed by Crick's ideas, because a common theoretical foundation for biology and chemistry held special promise for DuPont's future. The company excelled in molecular science. If biology were essentially molecules, then DuPont could excel in biology, too. The company's success with sulfonylurea herbicides encouraged Jefferson to move further into life sciences research.

In 1984, the 80th anniversary of the Experimental Station, a proud Jefferson joined CR&D director Richard Quisenberry at the famed Brandywine research site for the dedication of the new, $85 million Greenewalt Laboratory for health sciences. As Jefferson called on federal officials to remove biotechnology research from what he called "regulatory limbo," DuPont's four new life sciences research facilities at the Experimental Station, at Billerica, Massachusetts, at Glasgow, Delaware, and at the Stine-Haskell Laboratory in Newark, Delaware — all dedicated in 1984 — emphasized the company's hopes for research breakthroughs in biology, genetics and medicine. The next year DuPont opened its Electronics Development Center in Research Triangle Park, North Carolina, and entered a joint venture with N.V. Philips of the Netherlands to produce compact discs.

As DuPont's identity as a discovery company took shape, its steady global expansion assured that it would have a strong international component. Plants opened or expanded on a regular basis through the 1980s: Kevlar® and Hypalon® chlorosulfonated polyethylene at Maydown, Northern Ireland; Lycra® brand elastane in Brazil; automobile finishes in Venezuela; agricultural chemicals in Thailand; Mylar® PET

CLOCKWISE FROM TOP RIGHT:

Ludox® colloidal silica is used in the investment casting of golf club heads.

Crop protection chemicals are made in Bangpoo, Thailand.

DuPont joined N.V. Phillips in a 1985 venture to produce optical disks for data storage.

The *Gossamer Albatross,* which flew the English Channel, weighs only 55 pounds.

The Crawford H. Greenewalt Laboratory at the Experimental Station was opened in 1984 for biotechnology research.

The plant at Maydown, Londonderry, Northern Ireland, was DuPont's first European production facility. Construction started in 1958 for a neoprene plant. During the 1960s, Maydown began producing Hylene® PPDI thermoplastic resins as well as a variety of plastics and Orlon® and Lycra® fibers.

The Haskell Laboratory for Health and Environmental Sciences — one of the oldest corporate toxicology labs in the world — began conducting studies for other companies in 2001. The lab shares a campus near Newark, Delaware, with Stine Laboratory.

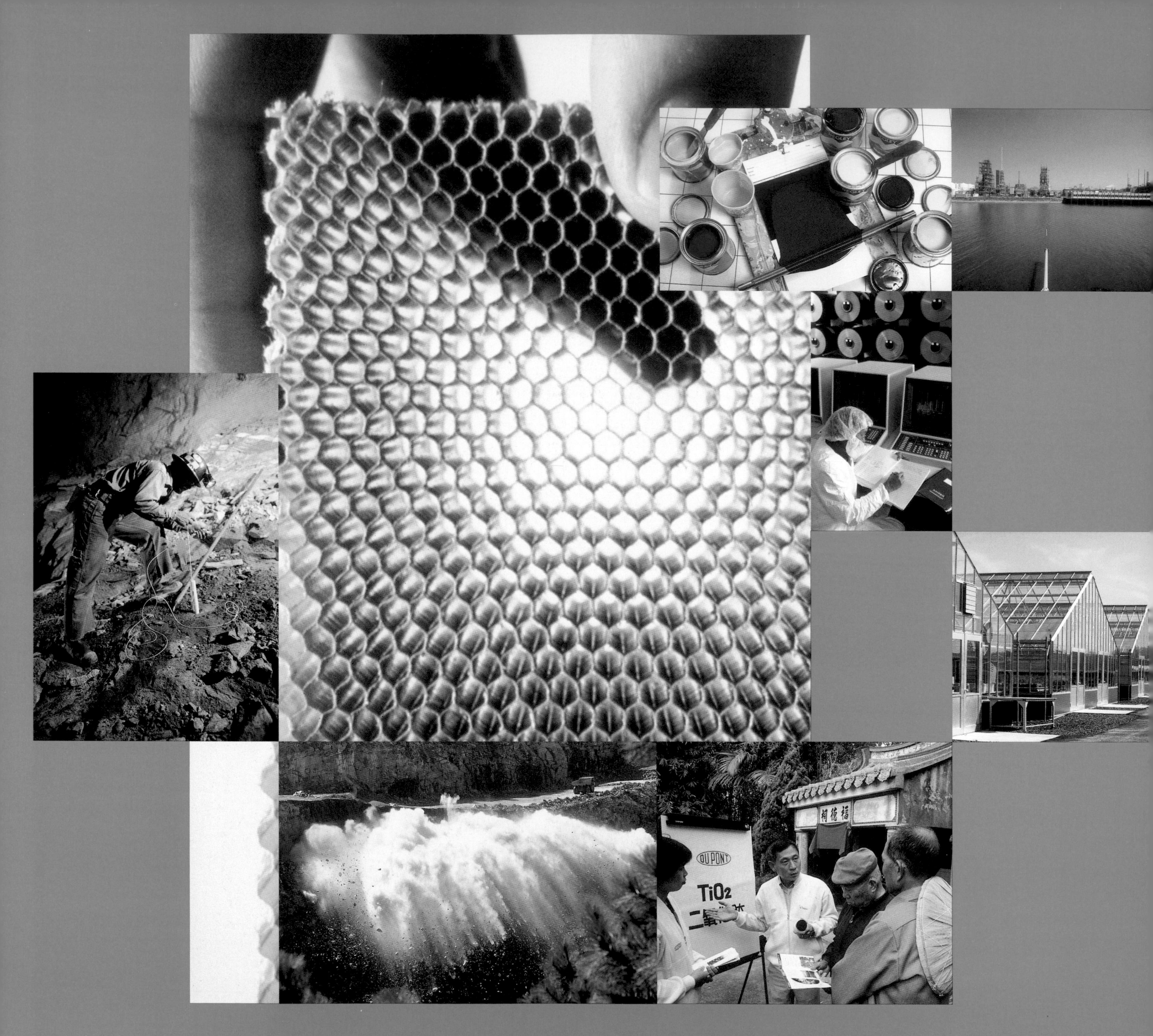
DU PONT
TiO2

polyester film and Tyvek® brand protective material in Luxembourg; Teflon® fluoropolymer in the Netherlands and Japan; carpet fibers in Germany; Glean® herbicides in Australia; and electronics in Mexico. In 1986 DuPont's titanium dioxide white pigment, Ti-Pure®, marked its third straight year of record sales, spurring plans for new TiO_2 plants in the Far East and South America. By the end of the decade, DuPont had become the United States' seventh largest exporter. DuPont operations in 40 countries meant that 45 percent of the company's total annual sales now occurred outside the United States.

In 1982 DuPont researchers announced their discovery of a new way to make tougher polymers, thereby extending applications for the company's engineering plastics like Zytel® nylon resin and Delrin® acetal resin as well as strengthening its existing 50 percent share of the U.S. automobile finishes market. By 1988 DuPont plants in 12 countries around the globe were producing engineering polymers, and plants in 10 countries produced automobile coatings. In 1983 DuPont introduced a new resin, Selar®, whose resistance to hydrocarbon solvents like gasoline made it an ideal liner for gas tanks and a replacement option for containers once made of glass or metal. Though Selar® barrier resin was manufactured only in the United States, DuPont produced other ethylene polymers, such as Surlyn® ionomer resin, in the Netherlands and Japan.

DuPont's global expansion through the 1970s and 1980s was part of a larger business strategy to find new markets while lowering production costs. Payroll was only one such cost, but at a company like DuPont it was a big one. In 1985 Vice President and Chief Operating Officer Richard Heckert appeared on videotape at scores of plants to offer an early retirement package to DuPont employees, many of whom, he said, had been hired in the 1970s for a boom that never came. The Executive Committee thought several thousand employees at most might respond, although the terms of the package were very generous. But 11,200 DuPonters — fully 10 percent of the company's U.S. workforce — took advantage of the $200 million plan and retired that May.

A year later CEO Edward Jefferson reached DuPont's traditional mandatory retirement age of 65, and DuPont's Board of Directors turned the office over to Richard Heckert. Heckert felt pressure from investors like the Bronfmans to trim DuPont's operations from the top down, as well as pressures from employees to preserve traditional ways of managing the company. DuPont had sold 20 businesses, including several Conoco assets, since 1981, and had discontinued many product lines, such as housepaints, dyes, antifreeze and car wax. Perhaps in a nod to Edgar Bronfman, who regarded the company as "extremely overstaffed" and its bureaucracy "mind-boggling,"[17] DuPont's Board of Directors pared the Executive Committee from nine to six members in 1986 — a minor adjustment that was matched by continued slow attrition in employee ranks of about 2 percent annually and by further curtailments of some long-standing operations.

Heckert, who had started his chemistry career at the Manhattan Project's Oak Ridge nuclear facility during a stint in the Army, had the right credentials to execute DuPont's withdrawal from the Savannah River nuclear plant in March 1989. DuPont's chemical engineers had designed Savannah River in the early 1950s using expertise they acquired at Hanford, Washington, during World War II, and DuPont personnel had operated the plant at cost for the U.S. government for nearly 40 years. But now the government was reconsidering its long-standing agreement to protect DuPont from lawsuits arising from the company's operation of the plant, and DuPont decided that the risks of continuing were too great. Other companies, specialists in nuclear engineering, could better manage Savannah River in the future, including its complicated nuclear waste disposal problems.

CLOCKWISE FROM TOP CENTER:

Honeycomb core structures of Nomex® combine exceptional strength, light weight and thermal protection in aircraft interiors, for example.

Ti-Pure® titanium dioxide pigments impart whiteness, brightness and opacity when incorporated into coatings.

The Dordrecht, Netherlands, plant manufactures Lycra® and Delrin®, among other products.

Mylar® polyester film became the primary substrate for video and audio recording tape.

Agricultural research is conducted at Stine Laboratory near Newark, Delaware.

As part of the company's entry into new communities, DuPont representatives meet with local residents, such as this 1990 gathering in Taiwan before the opening of a new Ti-Pure® plant.

Explosives — the company's sole product for its first 100 years — were phased out in 1988.

Dynamite was last made by DuPont in 1977, giving way to safer Tovex® water gel explosives.

TOP AND MIDDLE: **DuPont developed Dymel® aerosol propellants — for applications such as metered dose inhalers — and Suva® refrigerants as environmentally safe replacements for CFCs.**

BOTTOM: **DuPont established the Experimental Station in 1903 near Wilmington, Delaware, to conduct and promote scientific research as a major platform for industrial growth.**

RIGHT: **Communication cables are insulated with Teflon® FEP fluoropolymer resin, invented and developed in the 1950s by Manville Bro.**

In the late 1980s DuPont's environmental concerns moved increasingly to the forefront of business operations, especially its production of chlorofluorocarbons (CFCs) like Freon® refrigerants. CFC refrigerants, aerosol propellants and insulation materials were the most ubiquitous group of chemicals in commercial production. But in 1974 two University of California chemists published data linking CFCs to the depletion of the earth's protective ozone layer. The EPA had pressed forward with a mandatory phaseout of CFCs, but DuPont and the rest of the chemical industry had argued for further, definitive scientific proof of CFCs' harmful effects. The stakes were enormous, for DuPont produced half of all the CFCs sold in the United States and 25 percent of supplies worldwide. Consumers depended on these chemicals for their air conditioners, refrigerators and aerosols, and as yet there were no practical replacements.

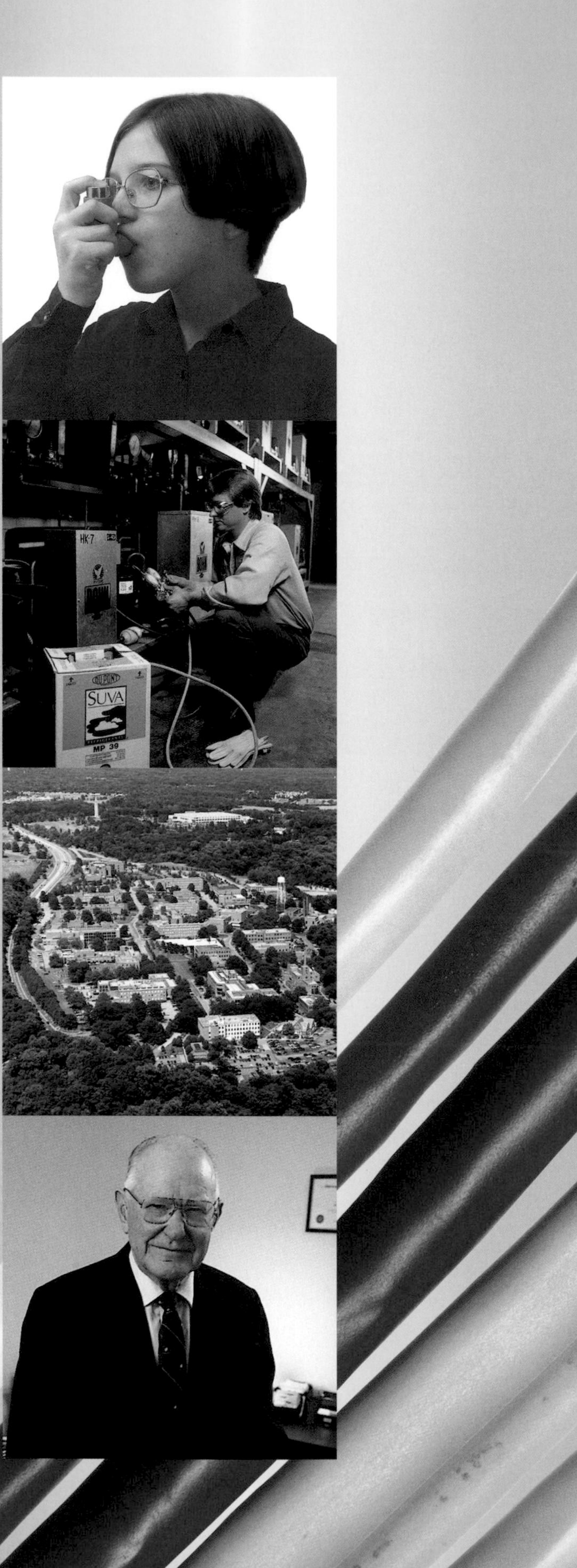

While some DuPont scientists worked to find CFC substitutes, others evaluated the data connecting CFCs to ozone depletion. Through the 1980s Central Research and Development's David Filkin and his colleagues applied new mathematical models that they shared with the National Aeronautics and Space Administration, helping that agency reach an important conclusion. On March 15, 1988, NASA scientists announced that CFCs were indeed depleting the ozone layer. These findings provided a sound scientific basis for DuPont's decision, reached just 72 hours after NASA's announcement, to cease CFC production by the turn of the century. This action put DuPont ahead of the industry timetable established by international agreement the previous year in the Montreal Protocol.

For scientist-executives like Richard Heckert, the CFC phaseout was a correct if difficult decision. "Chemists are people too," he said, while acknowledging that the biggest challenge facing the chemical industry was "convincing the public that we are their friends and not their enemies." If DuPont had suddenly stopped production of

In 1987 retired chemist Charles J. Pedersen was awarded the Nobel Prize, the first in the company's history. Pedersen spent his earliest years in Korea, but at the age of eight he enrolled in boarding schools in Japan. From there he went to the University of Dayton in Ohio, graduated with a degree in chemical engineering, then earned his master's degree in organic chemistry from MIT and went to work for DuPont at the Chambers Works' Jackson Lab in 1927. Pedersen was promoted to research associate, DuPont's highest researcher rank, in 1946. Eleven years later he transferred to the Experimental Station, where he began a line of investigation that resulted in a lengthy 1967 article in the *Journal of the American Chemical Society* that his colleagues jokingly called "the blockbuster." The article announced Pedersen's discovery of a new class of chemical compounds he called "crown ethers." The ring- or crown-shaped molecules of these crystal compounds had the unique ability to trap and hold certain ions. Their discovery opened new avenues for research, such as how the body recognizes certain proteins as foods and breaks them down into usable components for building new cells and tissues. Pedersen retired two years later at age 65 and spent the next 20 years happily fishing, gardening and writing poetry. Then, on the morning of October 14, 1987, he received a surprise call from the Nobel Foundation telling him of his award. "I was flabbergasted," he said.

World
Environment
Make your career count.
Making a difference every day.
CAREER OPPORTUNITIES:
• Physical Sciences
• Chemical Sciences
• Life Sciences
• Engineering
• Business
• Information Technology
• Student Employment
WORK/LIFE PROGRAMS:
• Family Leave
• Family Resource Programs
• Flexible Work Schedules
www.careers.dupont.com
Fax your resume to:
800-978-9774
Discover the DuPont Difference.
DUPONT

CFCs, Heckert explained, there would have been an outcry against the company for cutting off such functional products.[18] The search for suitable substitutes took time, but by the end of 1989 DuPont had filed 20 patents for non-CFC refrigerants. In 1990 the company started commercial production of Suva® refrigerants and Dymel® propellant.

As DuPont's chief spokesperson in the latter half of the 1980s, Richard Heckert strove for balance and perspective in a litigious and acquisitive era. He spoke not just for himself but for DuPont's traditions when he observed that Americans had become preoccupied with "acquiring wealth by rearranging it, instead of acquiring wealth by creating it." Shareholders deserved expert management of their assets, but they also owed society the benefit of a long-term view that allowed industry to create real value, not just short-term and shortsighted profits.[19] The relatively prosperous years after mid-decade relieved some of the pressure on Heckert to raise profits by cutting and paring DuPont's structure. The company's net income rose every year between 1986 and 1989, as did its net rate of return on investment, which very nearly reached the company's goal of 16 percent in 1989 — nearly double what it had been in the recession year of 1982.

In April 1989, 65-year-old Heckert retired, and only a few months later DuPont's new CEO, industrial engineer Edgar S. Woolard Jr., was happy to report to stockholders that the company's return on investment had at last reached the long-desired goal of 16 percent. Common stock earnings in excess of $9.00 per share justified a 3-for-1 stock split in December. Woolard, who started at the Kinston, North Carolina, Dacron® polyester plant in 1957, described his career at DuPont as advancing through 15 years of "golden age" prosperity, then 15 more of company "soul searching" and reorganization. Woolard's department, Fibers, had been hit hard in the energy crisis of the 1970s, and he subsequently served a stint in Corporate Plans, trying to help the company respond to the new conditions. He concluded that DuPont, like other large U.S. corporations, was in a new era that demanded a clear focus on market value. But where others saw woe and decline, Woolard saw that the company could actually function better with fewer layers of management and more individual accountability.

Though still a newcomer to the Executive Committee in 1983, Woolard was the first member of the group to take direct line responsibilities for actual industrial operations — in agricultural chemicals and in medical and photo products. By 1986, each of the Executive Committee's six members had assumed direct line responsibilities. Woolard was the latest in a line of DuPont executives going back to the 1960s who had questioned the role of the Executive Committee. But now, reformers like Woolard challenged the committee's existence, not merely its role. In the fall of 1990 DuPont's Board of Directors decided to abolish the venerable committee that Pierre had established 87 years earlier but that now, as Woolard put it, "sat up on the ninth floor of the DuPont building and read reports."[20] In its stead Woolard formed a five-person Office of the Chairman. Now only one person, a vice president, stood between Woolard and the line executive heading a particular business, like nylon, where before there had been three intervening managers.

Woolard also stepped up the pace of change in many of DuPont's employee policies. The company expanded its support for child care, flexible schedules and maternity/paternity leave to working parents. In 1990 the magazine *Working Mother* ranked DuPont in the top 10 of 75 U.S. companies considered the best for women with full-time jobs. Woolard reinvigorated DuPont's commitment to diversity, insisting that he was not out to cause a revolution but simply "to use the enormous talent we have — all of it."[21] DuPont stretched those talents across a widening cost gap in the early 1990s. Like many others in America's burgeoning information age, DuPont employees were learning more and doing more — and learning to do more with less.

OPPOSITE **In 1987, Edgar S. Woolard, then DuPont president, accepted on behalf of the company the Gold Medal of the World Environment Center for "international corporate environmental achievement." Presenting the award was Lee M. Thomas, administrator of the U.S. Environmental Protection Agency.**

In the early 1990s DuPont initiated programs to develop a more diverse work force.

BELOW **DuPont, a world leader in industrial safety performance, also is a world leader in providing safety training to first responders and hazardous materials workers.**

CHAPTER 9

THE PATH TO SUSTAINABLE GROWTH

FAR RIGHT and BACKGROUND: DuPont began producing Nomex® in Asturias, Spain, in 1993 and has since added facilities for Sontara®, Corian®, fungicides and tetrahydrofuran. The site is an award-winning showplace for corporate environmental stewardship.

RIGHT: Stacey J. Mobley was named senior vice president, chief administrative officer and general counsel in 1999.

BELOW: Edgar S. Woolard Jr. became DuPont chairman and CEO in 1989. He retired as CEO in 1995 and as chairman in 1997.

By 1991 CEO Edgar S. Woolard Jr. had made good on his promise to "shake up the bureaucracy" at DuPont, eliminating overlaps and redundancies in the company's management. A newly streamlined organization allowed employees to respond more quickly and flexibly to customers' needs. But the 1990–1991 recession added new urgency to this revitalization effort, when rising costs forced the company's many customers to look hard at DuPont's prices. "We don't want to go elsewhere," Woolard recalled hearing customers say, "but we must or we won't survive." One long-time polyester customer, who had routinely bought millions of dollars' worth of DuPont products, was especially adamant. "It's over!" he exclaimed to Woolard in frustration, then added more calmly, "I'll give you a year — six months — we've got to find a way to keep working together."[1] So there it was. From a customer's point of view, DuPont's streamlining hadn't gone far enough or fast enough. Now the company geared itself for a top-to-bottom overhaul.

Woolard's management team set an ambitious goal: to reduce fixed costs by $1 billion within two years. By the end of 1992 the company's 125,000 employees, 8,000 fewer than in 1991, had achieved 79 percent of that goal. But despite their efforts, annual sales dropped 2 percent in 1992, and the company's net earnings, at $975 million, had dipped below the $1 billion mark for the first time since 1979. The following year management announced further plant closings and layoffs; trimmed its own ranks to 60 percent of 1991 levels; and undertook a two-year, $500 million cost reduction program in the company's European operations. Woolard had worked closely with many of the employees who were laid off and knew that they had believed their work was essential to the company. He frankly called 1991–1994 "the low point for me, personally, in my career."[2] Senior Vice President for External Affairs Stacey J. Mobley thought Woolard had been "Churchillian" through the prolonged crisis — boosting spirits with an unflagging confidence in ultimate victory — but Mobley also saw the private side of DuPont's CEO. "He was the brunt of a *lot* of personal attacks, from a *lot* of friends he had grown up in the company with, and it wore on him," said Mobley. "It very much wore on him."[3]

In the early 1990s many Americans struggled to accept the new reality of employment in a global economy. For many years America's economic strength had meant widespread job security for most workers. Now, however, the pressures of worldwide competition exposed them increasingly to business risks.[4] As Woolard put it in 1993, "Security comes only from providing superior value to customers," which required new levels of resourcefulness, initiative and creativity from every individual in the company.[5] DuPont employees around the world shared the sacrifices necessary for the company's survival in the 1990s, while the company's manufacturing activities often moved closer to markets around

BACKGROUND and NEAR RIGHT: **Florida's Broward Center for the Performing Arts is protected from hurricanes, vandals and noise by windows with SentryGlas® Plus ionoplast interlayer.**

FAR RIGHT: **Since 1986 Stainmaster® carpets have provided consumers with superior stain-resistant flooring.**

LOWER RIGHT: **DuPont sleep products include Comforel® pillows, comforters and mattress pads.**

the globe. In 1993 the company opened fibers and elastomers plants in Asturias, Spain, and constructed plants to make Lycra® and Ti-Pure® titanium dioxide pigment in Singapore, while planning for 20 new operations in Asia. At home, DuPont strengthened some traditionally successful product lines. It launched three new brands of carpet fiber; announced SentryGlas® composites with Butacite® sheeting for new safety windows, the first window product to meet new safety codes for resisting hurricane damage; and introduced its Synchrony® STS® package of sulfonylurea herbicide and soybean seeds to farmers in the Midwest. The company's 30-year-old pharmaceuticals business, boosted by a 1991 joint venture with Merck & Co., offered highly effective products to cardiac and stroke patients and helped the DuPont Merck Pharmaceutical Company post strong earnings through the mid-1990s.

Recalling the comments of his dissatisfied polyester customer two years earlier, Woolard turned the company's five major business segments — Petroleum, Chemicals, Polymers, Fibers and Diversified Businesses — upside down in 1993 and shook out 20 new "Strategic Business Units" (SBUs) that he hoped would be lean, customer-focused and flexible. DuPont CEO Charles McCoy had introduced this concept to the company back in 1968 with his "Profit Centers," but now Woolard stretched the idea to new dimensions, holding the SBUs directly accountable to him and a small number of senior vice presidents. He also instituted a new stock option plan for employees, giving everyone a direct, personal stake in the company's future.

DuPont's administrative and support services also were reshaped to fit the company's leaner, more flexible business model. Nowhere was the change more dramatic than in DuPont's research activities. Though not as visible as the company's many products, DuPont's research had evolved apace with the rest of the company. By the early 1990s it constituted an extensive, worldwide enterprise, with Central Research and Development's (CR&D) basic science work at the Experimental Station and Chestnut Run accounting for just 10 percent of the total R&D budget of nearly $1.5 billion. The vast majority of the company's research was conducted in laboratories organized, staffed and funded by the industrial departments. Over the years those research and development efforts had become very large, with more emphasis on development than research. They focused mainly on problems with direct relevance, or at least a substantial promise of it, to the departments' products and markets. Meanwhile CR&D remained essentially what Charles Stine had envisioned in the late 1920s — an intellectual powerhouse of original scientific inquiry loosely bound by the promise of future profitability.

The 1990–1991 recession exacerbated the long-simmering, sometimes creative, but often chafing tension between CR&D, also called "central research," and businesses operating on the front lines of change. Industrial departments trimmed their own research operations and then were themselves transformed by Woolard's team into leaner SBUs. If they were going to have to do more with less, DuPont's business managers reasoned, they were going to need a more productive relationship with central research. "There were brilliant people in CR&D who could do wonderful things if you could only figure out how to link them with the company," recalled Al MacLachlan, then senior vice president for R&D. His task at the time, he said bluntly, was "either link 'em or dump 'em. And we linked 'em."[6]

That accomplishment came only after months of careful diplomacy. On the one hand, scientists and managers in the company's business units needed assurance that corporate researchers would focus on issues with tangible benefit to customers. On the other, MacLachlan recognized, CR&D researchers could not be forced or cajoled into creative responses to businesses' needs. Like most people, they needed to feel appreciated, to be included as valuable members of a team. But valuing them for solutions to short-term problems could not be allowed to compromise their freedom to consider

17036
LIONS

LIQUID BEVERAGE POUCHES (LEFT) — AN IDEA DEVELOPED BY DUPONT CANADA — ARE AS PREVALENT IN MEXICO AS SOLID CONTAINERS. IN THE UNITED STATES, HUNDREDS OF SCHOOLS HAVE SWITCHED TO MINI-SIP® POUCHES FOR MILK AND JUICE. THE DUPONT RHYTHM® (REMEMBER HOW YOU TREAT HAZARDOUS MATERIALS) PROGRAM (FAR LEFT) HAS BECOME AN INDUSTRY MODEL FOR TRAINING OFF-SITE MANUFACTURERS' EMPLOYEES IN PROVEN METHODS OF HANDLING AND TRANSPORTING CHEMICALS. FOR THE ROOF OF PONTIAC STADIUM NEAR DETROIT (BOTTOM LEFT) AND THE ROOF OF THE MAIN TERMINAL AT DENVER INTERNATIONAL AIRPORT (BELOW), ARCHITECTS CHOSE FIBERGLASS ROOFING FABRIC COATED WITH TEFLON® PTFE FLUOROCARBON RESIN. THE ROOF LETS IN PLENTY OF LIGHT, BUT ABLY KEEPS OUT WIND AND SNOW. UNDERFOOT AT THE AIRPORT, CARPET OF ANTRON® LEGACY NYLON RESISTS SOIL AND MAINTAINS ITS TEXTURE DESPITE THE POUNDING OF MILLIONS OF FEET.

novel, unusual, even unlikely approaches. There was a critical balance point that no one could measure precisely but that nonetheless had to be discerned and respected.

Over several months MacLachlan and Vice President for R&D Richard Quisenberry helped central and business unit researchers forge a new, rewarding relationship. Instead of competing for resources, they began to share equipment, such as an expensive nuclear magnetic resonance spectrometer, as well as personnel and ideas. They even shared the burden of reducing costs in the early 1990s. "They went at that with a vengeance," said MacLachlan. "We knocked about $100 million out of total R&D in a matter of a few months."[7]

Other DuPont support functions achieved similar results. For example, DuPont Legal reduced the number of outside law firms it used from 350 to 34, nearly halving the company's legal fees between 1994 and 1998. The aim of these and other changes was not simply to get smaller but to adapt to the turn-on-a-dime conditions that new communication technologies made possible — and that new competition made necessary. No longer would outside law firms pour time and money into litigation that did not merit the effort. Instead, they would work closely with DuPont Legal and with one another to process legal matters, including litigation, economically and efficiently for the company.

During the same period DuPont implemented its new legal approach, the company also faced its most significant set of product liability cases, the Benlate® litigation. Introduced in a wettable powder form in 1970, Benlate® had long been one of the company's most successful fungicides and was registered worldwide for many crops. In 1987 DuPont introduced an alternative, dry-flowable form (Benlate® 50 DF) that was recalled in 1989 and 1991 owing to the presence of the herbicide atrazine in some lots. The recalls generated hundreds of claims, and growers and their lawyers began blaming Benlate® 50 DF (even product free of atrazine) for a wide range of plant problems. DuPont initially paid many claims to maintain good customer relations, and at the same time initiated the most intensive investigation in the history of U.S. agriculture to determine whether Benlate® 50 DF could cause plant damage. When the testing could not duplicate the claimed plant injuries, the company declined to pay any further claims.

In the following decade, DuPont faced hundreds of Benlate® lawsuits. The litigation results were mixed. DuPont won cases before some courts, including a Florida administrative proceeding that found nothing wrong with the product. Other trials resulted in losses, including some for large amounts that reflected the runaway verdicts being rendered by the U.S. jury system in the 1990s. Ultimately, for business reasons, the company decided to stop selling Benlate® worldwide in 2001, even though there is still no credible scientific evidence demonstrating that Benlate® caused either the crop or health problems alleged in the lawsuits. This decision came as a disappointment to many growers, who continued to rely on Benlate® as a safe and effective product throughout the period of litigation.

In 1994, despite its worsening Benlate® headache, DuPont realized a net earnings gain of 65 percent over the previous year. Four years of sacrifice and restructuring had at last paid off. The company's employees had greatly surpassed the cost-savings goal of $1 billion that Woolard had set in 1991, trimming $2.5 billion in fixed costs in just over three years. Woolard's genial but iron resolve had helped sustain the company through a difficult period while embracing, not avoiding, challenges to reduce pollution and waste. In 1989 DuPont announced a new commitment to "corporate environmentalism," a term first coined by Woolard to describe the company's renewed sense of environmental stewardship. The catastrophe at Bhopal, India, in 1984; the Chernobyl nuclear reactor meltdown in the Ukraine in 1986; and the 11-million-gallon crude oil spill in Alaska's Prince

DuPont researcher Jim White came upon Tyvek® in 1955. It is now widely known as a construction wrap for homes and commercial buildings as well as a tough-to-tear envelope material. The product line has been extended frequently, now including such items as Tyvek® Sendables® for shipping gifts (background). The Smithsonian Institution (inset) uses nonabrasive Tyvek® to protect artifacts. Tyvek® also is used for sterile medical packaging and in graphic applications such as outdoor advertising banners.

Quick-drying CoolMax® high performance fabric keeps the wearer dry and comfortable through a range of climates and activities. CoolMax® moves sweat away from the body to the outer layer of the fabric, where it dries faster than any other fabric. When temperatures drop, Thermolite® performance insulations keep the wearer warm and active.

William Sound when the *Exxon Valdez* struck a reef in 1989 made corporate environmentalism more than smart public relations in the 1990s. "Manufacturers have been painted many colors in recent years," said Woolard soon after his appointment as CEO, but in the future they would have to be seen as all one color. "And that color had better be green."[8]

In May 1989, in his second week as CEO, Woolard had traveled to London to speak to the American Chamber of Commerce about industry and the environment. The goals he outlined for DuPont in that speech placed the company in the forefront of environmental change: 35 percent reduction in hazardous wastes by 1990, with at least 70 percent reduction by 2000; elimination of the heavy metal pigments used in making certain plastics; a polyester recycling venture; inclusion of community representatives in all major site planning activities relevant to public health or the environment; and inclusion of environmental performance — "both pro and con," said Woolard, pointedly — in determining executive pay. Additionally, Woolard pledged DuPont to set aside company-owned land around the world as nature preserves. DuPont would no longer just react to environmentalists' initiatives and government regulations; instead, the company itself would set new standards for others to follow around the globe. "As DuPont's chief executive," Woolard told his London audience, "I'm also DuPont's chief environmentalist."[9]

Woolard knew the London speech would be an important one, so he asked his oldest daughter to read it first, as she did many of his speeches. Her reaction to this one surprised him. "Dad," she said, "I have just one question — do you believe what you're saying?" She was not alone. Woolard later recalled that his speech had stunned many at DuPont who wondered "whether this guy was for real."[10] Admittedly, he had not

SPECIALIZED
EPIC
SPECIALIZED
Giro

OPPOSITE: **Commercialized in 1962, Lycra® brand elastane fiber capped two decades of research to produce a good synthetic elastomeric fiber. Lycra® was made first in Waynesboro, Virginia. The stretch and recovery properties of Lycra® provide fabrics and garments with comfort, fit and freedom of movement. Lycra® long ago moved from intimate apparel and swimwear to activewear, men's wear, denim, footwear and even Leather with Lycra®.**

BELOW: **John A. Krol was chairman of the company from 1997 through 1998.**

consulted any scientists before announcing the new environmental commitments in London. When he returned to Wilmington he ran into strenuous objections from some of the company's leading scientists and business managers, who protested that they could not possibly meet the goals he had set for them. Their reaction rankled Woolard, who had a bedrock faith in DuPont's human talent and historical ability to overcome obstacles.

The skepticism of influential scientists within the company was worrisome, and Woolard met it head-on. Calling a large meeting at the Experimental Station, he explained the company's new commitments to the assembled researchers, and gave doubters a positive but positively unyielding message. "I've got more confidence in you than you've got in yourself," he said. "Now just go do it."[11] To underscore the seriousness of his resolve, Woolard gave Paul Tebo, vice president of Safety, Health and Environment, a tough assignment: "Go find the most environmentally unsound plant we have and give them a year to change; and if they don't I'm going to shut 'em down. We're going to make this happen."[12] A year later the plant at Beaumont, Texas, boasted a two-thirds reduction in emissions as well as improved production yield and a net savings of $1 million.[13] DuPonters there had done more than find better ways to dispose of waste. They had actually reduced the amount of waste by changing the manufacturing process itself.

THE GOAL IS "0"

Soon and for the first time, DuPont dared to mention "zero emissions" as a possible company goal. Paul Tebo approached Woolard on behalf of a group of the company's influential managers who wanted DuPont to commit publicly to "zero waste and zero emissions" by the turn of the century. Woolard asked for the commitment in writing. When Tebo returned with 65 signatures, Woolard went forward with the announcement. Environmentalism had proved to be, as Woolard described it, "the most empowering and unifying initiative I've ever seen in this company," extending to employees throughout the organization.[14]

Woolard described a trip he made during this period with the company's board of directors to the nearby Chambers Works plant in New Jersey. It was the directors' annual visit to a DuPont facility, but this time, there was a change in the usual protocol. As the board members wound their way through the plant, an equipment operator, not a plant manager, explained to them a particular chemical production process. The operator enthusiastically described how he had recently rerouted some pipes in a way that reduced both the plant's costs and its emissions. "This is so great!" exclaimed Woolard, "I'm so proud of you!" As everyone beamed, a thought suddenly flashed in Woolard's head. With a glance at the directors, he asked the operator, "How long have you had that idea?" "About ten years," the man replied, making Woolard's point as well as anyone could have — that people can think creatively if they're empowered and encouraged to do so. "It wasn't a huge thing," Woolard admitted, "maybe saved $100,000 a year; but what you want is a thousand people thinking like that."[15]

In March 1995, just when DuPont had completed most of its restructuring and when profits were climbing, the Bronfmans, owners of The Seagram Company Ltd. and members of DuPont's board since 1981, offered to sell back to DuPont their 156 million shares of DuPont common stock for $8.8 billion in cash so that Edgar Bronfman Jr. could invest in the entertainment field. Woolard accepted the offer, partly because reducing the number of shares increased their value, but also because there was little point in holding onto an investor whose heart was elsewhere. Borrowing the cash required to accomplish the buyback doubled DuPont's indebtedness to $16.2 billion. That year the company sold two medical products businesses and $1.7 billion of new stock to help reduce its debt, and undertook a flurry of international ventures to help ensure future earnings — nylon in Taiwan, India, Japan, Brazil and Mexico; Lycra® and Mylar® in China; and Zytel® in a new $100 million plant in Singapore. The

PIONEER
8379

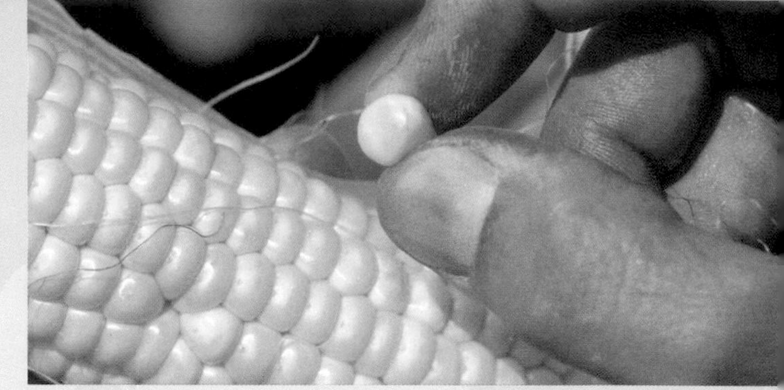

company also began construction of a Lycra® intermediates plant in Korea as well as a plant for its spunbonded fiber, Sontara®, in Spain, and at home announced a joint venture with Dow Chemical, DuPont Dow Elastomers LLC.

Between 1989 and 1995 DuPont had changed dramatically the way it *did* business, but it had not changed the businesses themselves very much.[16] When John A. "Jack" Krol succeeded Woolard as DuPont's new CEO on December 31, 1995, the company entered a new and important phase of its transition from a 20th century chemical company to a 21st century global science company. Restructuring had eliminated inefficiency and increased the company's productivity. Now Krol sought to realize the growth potential that those years of restructuring had created. Krol had joined DuPont as a chemist in 1963 and later held senior management positions in the company's Agricultural Products and Fibers businesses. These diverse experiences gave him a special appreciation of DuPont's potential in both biology and chemistry. As CEO he looked for opportunities to expand DuPont's activities in the life sciences while also strengthening the company's traditional product lines.

DuPont researchers found new molecular bridges connecting chemistry with biology. Late in 1995 DuPont launched a new business, DuPont Qualicon, based directly on research discoveries in molecular biology. DuPont Qualicon marketed the company's RiboPrinter® and Bax® systems, which were able to identify all existing bacteria and several other forms of food contaminants by scanning genetic information such as DNA "fingerprints." After only one year, these systems were being used by laboratories in nine countries, and in 1999, the U.S. Centers for Disease Control and Prevention adopted the RiboPrinter® to detect harmful bacteria in food.

In 1996 DuPont and a biotechnology firm, Genencor International, patented an *E. coli* microorganism genetically engineered to produce 1,3 propanediol, or PDO, a key polyester ingredient. PDO was usually obtained from petroleum, but the new microorganism could make PDO from readily available carbohydrate sources like corn sugar. It was a remarkable achievement, a real miracle of science that also posed new challenges to existing legal concepts of intellectual property, such as the question of whether new life forms can be patented.

Net earnings reached a new record high of $3.6 billion in 1996, while DuPont's 3,700 scientists worked hard in laboratories around the world to expand the technologies supporting the company's global growth. The Versipol® (versatile polymerization) process for making polyethylene at low pressures reduced the costs of manufacture and created a greater variety of these materials for packaging and industrial uses. DuPont's 1995 patent application for Versipol® involved 528 separate claims, making it the largest patent ever filed by the company up to that time. Subsequent licensing of Versipol® technology gave DuPont a key position in a fiercely competitive segment of the global plastics market.

In September 1997 DuPont purchased a 20 percent interest in Pioneer Hi-Bred International Inc. of Des Moines, Iowa, the nation's leading producer of corn and soybean seeds. DuPont also established a joint research venture with that company, Optimum Quality Grains LLC, later DuPont Specialty Grains, to develop products for the animal feed market.[17] Three months later the company bought Protein Technologies International Inc., a soy protein producer, from Ralston Purina. These decisions fulfilled Krol's aim to refocus DuPont on growth in the life sciences, but an additional major purchase that year proved disappointing. DuPont bought ICI's polyester intermediates and resins businesses and its TiO_2 pigment plants as a package, fully expecting that U.S. regulators would approve the deal. However, when federal officials disallowed the TiO_2 portion of the agreement on antimonopoly grounds, a surprised DuPont resolved to make the most of the remaining polyester portion.

Increasingly, observers inside and outside the company worried about DuPont's substantial investment in traditional stalwarts like chemicals, polymers and textile

The October 1999 acquisition of Pioneer Hi-Bred International Inc. marked a major step in DuPont's overall strategy to integrate agricultural biology into the company's science and technology base. Founded in Des Moines, Iowa, in 1926 by Henry Wallace, Pioneer was the first firm to engage in the production and marketing of commercial hybrid seed corn. Pioneer Hi-Bred International continues to develop new foods with higher nutritional value while cutting environmental waste, and is the leading developer and integrator of agricultural technology.

BACKGROUND: **Research engineer Mark Armstrong was part of a 1990s thrust into holographic technology for brighter electronic display screens.**

RIGHT TOP: **SilverStone® non-stick coating is now used on kitchen tools and accessories.**

FAR RIGHT TOP: **DuPont began producing Tynex® nylon monofilament for toothbrushes in 1995 in India.**

FAR RIGHT CENTER: **Zytel® ST super tough nylon provides stiffness, high impact strength and resilience to fire hose nozzles.**

RIGHT BOTTOM: **Italian bottle manufacturers use Melinar® Laser+ resin for 2-liter Coke® bottles.**

BELOW: **Charles O. "Chad" Holliday Jr. became DuPont chairman in January 1999.**

fibers, for which global competition was especially sharp, technology aging, and growth potentials limited. Krol recognized that globalization was part of the solution to the problems that global competition had created. Accordingly, DuPont sought to produce and market products wherever they had not yet become widely used commodities. For example, DuPont Hongji Films Foshan Co. Ltd., a joint venture in China's Guangdong province, yielded an annual output of 22,000 tons of polyester film in 1996, making it China's largest domestic supplier of that product. Similarly, DuPont formed a subsidiary in Santiago, Chile, to make TiO_2 pigment, Lycra®, Dacron®, nylon and polyester. But a worldwide presence only set the stage for the complex business strategy that DuPont needed to realize its full potential as a science-based, innovative competitor.

In October 1997 DuPont's board of directors named 49-year-old industrial engineer Charles O. (Chad) Holliday Jr. to succeed Jack Krol as DuPont's CEO in February 1998, while Krol, who became chairman of the board, moved into a long-intended retirement. Since joining DuPont in 1970, Holliday had seen service in virtually every aspect of the company, including seven years based in Tokyo as head of DuPont Asia Pacific. His enthusiasm for DuPont's future was palpable, but so was his regard for the company's legacy. "Dick Heckert told me once," he said, "'This company lives by the letter of its contracts, and also the intent.' That really struck home. In the end we may leave a dollar on the table in the short term but in the long run we'll make a lot more."[18]

Holliday moved quickly to enhance DuPont's global market strengths. He sold off company assets worth $1.2 billion, added to that figure $1.7 billion from DuPont's record $4.1 billion net earnings that year, and borrowed the difference to help pay for $10.2 billion worth of acquisitions. DuPont sold its global hydrogen peroxide business, printing and publishing businesses, and remaining 50 percent interest in Consol for a total of $865 million, then turned its attention

8th
continent
soymilk
original

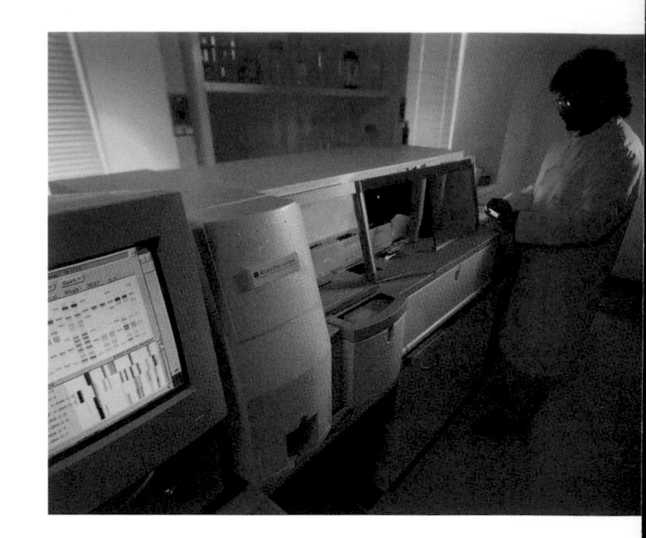

to Merck, which now wished to leave its pharmaceuticals joint venture with DuPont. Certain DuPont Merck products, like Cozaar® blood pressure medication and Coumadin® anticoagulant, had proven both effective and profitable, and DuPont's new Sustiva™ would soon receive final U.S. Food and Drug Administration (FDA) approval as a treatment for human immunodeficiency virus (HIV). DuPont regarded Merck's marketing experience as a good fit with DuPont's research strengths, but Merck wanted to go it alone. In July 1998 DuPont bought out Merck's 50 percent share of the DuPont Merck Pharmaceutical Company for a cash payment of $2.6 billion, established DuPont Pharmaceuticals and began searching for another, suitable alliance.

The need for investment capital as well as the ever-present risks of the oil business soon prompted DuPont to reassess its Conoco asset. Oil prices dropped sharply in 1998 and cut Conoco's earnings by 38 percent, contributing to a drop in DuPont stock value from an all-time high of $84 per share in May to $53 at year's end. Oil prices had never reached the sustained high levels many had anticipated when DuPont bought Conoco in 1981, and the 17-year-old decision had remained under a cloud. Now that cloud was casting a shadow on DuPont's horizon. Conoco's earnings had often helped DuPont through hard times, but the costs of oil exploration were steep and the payoffs unpredictable. Moreover, the corporate cultures of Conoco and DuPont had never fully blended. Sprung from wildcat origins in Ponca City, Oklahoma, a century earlier, Conoco had chafed in even the loose restraints that DuPont imposed. "They had their pride and their history," said DuPont's Stacey Mobley, "and they were not about to give it up, even after being acquired by DuPont."[19] On May 11, 1998, DuPont announced it would set the oil giant free, thus liberating itself for new ventures.

In October DuPont parted with 30 percent of Conoco in the largest independent public stock offering in U.S. history, netting the parent company $4.2 billion. Through the end of 1998 and into 1999, DuPont disposed of the

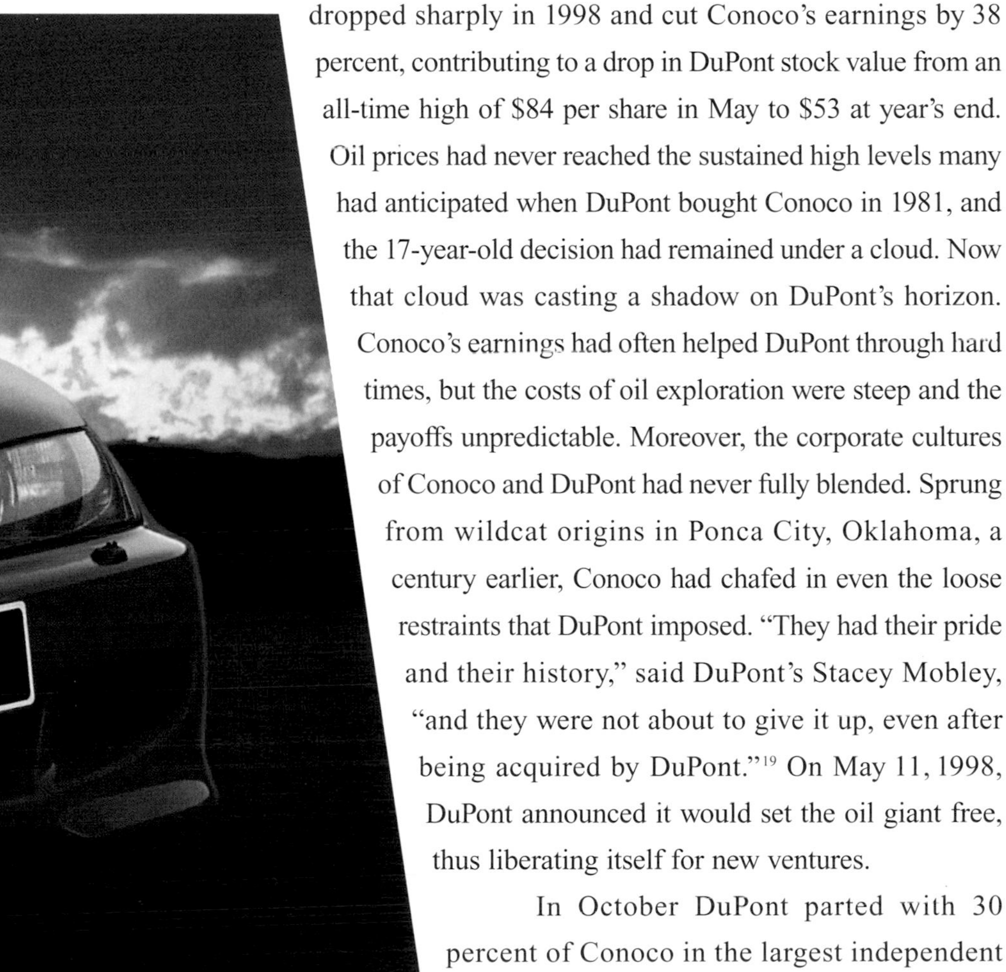

NEAR LEFT: 8th Continent — a joint venture between DuPont and General Mills — introduced three flavors of soymilk in 2001 containing Solae™ brand soy protein from DuPont Protein Technologies.

FAR LEFT: 1995 Harley-Davidson motorcycles took to the road with some 45 parts of Zytel® ST nylon resin.

CENTER LEFT: DuPont subsidiary Qualicon offers products and services utilizing its expertise in genetic technology and molecular biology to improve quality and safety in the food and health markets. Qualicon developed the RiboPrinter® Microbial Characterization System for fingerprinting bacterial DNA as well as the BAX® System, an automated, genetics-based platform for pathogen detection. The business also provides global food safety consulting services.

BOTTOM: The BMW Z3 roadster won environmental plaudits for its water-based paints, many from DuPont Herberts Automotive Systems.

PHILADELPHIA
CREAM CHEESE
Kellogg's
CORN FLAKES
Como de la familia!
STANLEY
Reduced Fat
RITZ
Lipton
Pomi
STRAINED TOMATOES
FAT FREE
Newtons
RASPBERRY
ARMOUR
Nestle
DUPONT

remaining 70 percent by exchanging Conoco shares for DuPont shares, adding an additional $7.5 billion to its earnings from the divestiture. DuPont quickly plowed this money into some of its most promising fields. In February 1999, for example, the company purchased the European automobile coatings firm Herberts from Germany's Hoechst A.G. for $1.7 billion, making DuPont the largest auto coatings supplier in the world.

Just a few weeks later, DuPont announced it would purchase the remaining 80 percent of Pioneer Hi-Bred International for $7.7 billion in cash and stock. Pioneer would remain in Des Moines as a wholly owned DuPont subsidiary. Some industry analysts considered Pioneer's purchase price steep, but Holliday's response reflected his experience, as well as that of Woolard and Krol before him, in the fast-paced 1990s: "Speed. Speed. Speed," he said plainly.[20] DuPont was big but it wasn't slow, and Holliday was going to make sure the company was out in front of the cresting biotechnology wave. In its 73-year history Pioneer had introduced nearly 100 new corn hybrids, and in recent years had tested about half a million experimental samples of corn and soybeans annually; nearly 70 percent of its seeds contained genes altered to improve insect or herbicide resistance. Major developments emerging almost daily from Pioneer's labs and from other research labs made the agriculture and nutrition market intensely competitive. Participants like DuPont needed to be alert, entrepreneurial and daring.

In November 1999, a month after the Pioneer deal was completed, the FDA answered Protein Technologies' petition favorably and authorized advertising claims linking the use of soy foods to a reduced incidence of heart disease.

CENTER: **All photo copiers and laser printers use fluoropolymers in their fusing units. Fumio Inomae, Toyko, works in the DuPont Teflon® finishes business, which holds more than 80 percent of this market. The non-stick and high temperature resistance properties of Teflon® are essential to document printing.**

FAR LEFT TOP and BOTTOM: **Cyrel® digital imaging technology enables company's such as Kraft Foods to economically enhance package graphics.**

FAR LEFT CENTER: **DuPont packaging evolved from moistureproof cellophane to a broad range of packaging materials, including Clysar®, Melinex®, Mylar®, Selar® and Surlyn®.**

NOTABLE

> Americans throw away about 29 billion pounds of plastic every year, an average of 106 pounds for every individual. Plastics fill 20 percent of the available landfill space in America, so DuPont researchers are helping Americans recycle their plastic throwaways by developing new, degradable plastic products. Chemist Steve Hanson helped invent Biomax® in the mid-1990s, a new polyester whose weak points turn out to be its strengths. Hanson and his colleagues figured a way to insert weak monomer links in the polymer chains that make up a widely used polyester plastic. The links hold up well for the product's many uses, which include beverage containers and peanut butter jars, but when exposed to the elements, as in a landfill, the monomer links give way to water damage and are even consumed by microbes and earthworms. But Biomax® has strong points other than its degradability in landfills and its nutritional interest for microbes and worms. It can also be re-used in containers, or as coating for disposable paper plates and cups. Turf can be grown on it too, creating rolls of new grass that are lighter to carry and easier to roll than sod.

> DuPont opened a dye works in 1917 at its Chambers Works facility in Deepwater, New Jersey, beginning a six-decade enterprise in synthetic dye research and production. DuPont closed the dye works in 1980 but Chambers Works continued to manufacture chemical intermediates and other products. This production drew the attention of the Environmental Protection Agency as it focused on the environmental problems associated with chemical production. In a move that surprised advocates and critics alike, DuPont joined forces with the EPA in 1991 to produce a $1 million study that identified 15 ways to improve the facility's environmental record. The changes included installing an expensive, high-pressure water jet system to clean vessels previously cleaned with a solvent that needed to be incinerated. But despite their high initial cost, new processes like this were not only cleaner for the environment but also more cost effective in the long run. Change has continued at the New Jersey plant. Today it is home to the DuPont Environmental Treatment (DET) facility, the world's largest commercial and industrial wastewater treatment plant. DET accepts wastewater from other corporations in the chemical, metalworking and automotive industries, delivered daily on barges, tank trucks and railcars. The wastewater is treated, cleaned and discharged in quantities of up to 40 million gallons per day in methods compatible with EPA standards. The DET processes 85 percent of all industrial wastewater generated in New Jersey.

> DuPont scientists today are fighting crop-destroying fungi that many people assume disappeared long ago. For example, the infamous potato blight that caused widespread starvation and death in Ireland in the mid-1840s has returned. This time, however, it is attacking potato crops in Mexico and parts of South America. About 1.5 million Irish died of starvation in the mid-1840s when unusually wet weather allowed a fast-spreading fungus to ruin the potato harvest several years in a row. During the 1850s, German botanist Anton de Bary

> Biologist Scott Sebastian joined DuPont in 1984 with a very specific goal in mind: to find a soybean with a genetic resistance to DuPont's sulfonylurea herbicides. Such a crop would survive a spraying that killed all other plants competing with it for water and nutrients — weeds, in other words. Sebastian screened 8,000 soybean varieties but not one showed the slightest resistance to the weed killer. So he stepped up the search in DuPont's greenhouses and developed a novel screening procedure that could search millions of soybeans quickly. Within three months the natural genetic variation that occurs randomly in any species's gene pool yielded a plant with sulfonylurea resistance that was passed on to following generations of soybeans. Using standard methods of plant breeding, Sebastian developed the variety and patented it, the first patented soybean in history. In 1993 DuPont marketed the product as a combination of Sebastian's new soybean seed and the company's herbicide Reliance®, calling it Synchrony® STS® (sulfonylurea-tolerant soybeans).

identified the fungus *Phytophthora infestans* as the cause of the blight, and chemical means were soon devised to quash it. But *Phytophthera infestans* evolved treatment-resistant strains. Killing some new strains now requires up to 12 times the normal application of traditional fungicides. In recent years, however, farmers worldwide have used fungicides like Curzate® with great success, while the company's sulfonylurea herbicide Matrix® effectively controls the weeds that threaten potato crops.

> In the 1990s, DuPont developed several proprietary methods for recycling plastics. While research advances often went unnoticed by the general public, the celebrated Revolutionary War ship replica *HMS Rose* sailed into ports all over the country under a magnificent spread of recycled polyester "canvas." DuPont manufactured the 13,000 square feet of sails for the majestic ship, the largest active wooden tall ship in the world. Made from 126,000 recycled plastic bottles and plastic car fenders, the sails delighted and impressed tourists and schoolchildren alike from Fort Lauderdale to Seattle. DuPont used PET (polyethylene terephthalate) plastic for the sails, transformed from bottles and fenders into plastic pellets and resin and then into fiber yarn. The yarn was then woven into sailcloth. The ship served as a symbol for the possibilities of recycling and for DuPont's leadership role in the national recycling effort. DuPont's plastic sails will hit the big screen when the *HMS Rose* appears in the film version of Patrick O'Brian's *Master and Commander*, the first novel in a series about the early 19th century high seas adventures of Captain Jack Aubrey and his ship's surgeon.

> DuPont's extraordinary experience in industrial safety has enabled the company to market this knowledge as a valuable service called DuPont Safety Resources. For instance, applying this knowledge to an unusual number of "slips, trips and falls" at the Johnson Space Center near Houston, DuPont safety consultants reduced the mishap rate there by 77 percent between 1994 and 1999, saving NASA an estimated $2.08 million.

CENTER and INSET: **Amtrak's Acela Express took to the rails of America's Northeast Corridor in December 2000 with Tedlar® polyvinyl fluoride films protecting interior surfaces.**

FAR RIGHT TOP: **Nokia cellular phones contain high-density printed circuit boards made with Riston® photopolymer dry film resists.**

FAR RIGHT CENTER: **Uma Chowdhry (standing) is director of DuPont Engineering Technology, a consulting organization that develops marketable research ideas into full-scale products and technologies.**

FAR RIGHT BOTTOM: **DuPont extended its nonwovens line with Sontara® AC™ aircraft wipes for manufacturing and maintenance.**

MIDDLE BOTTOM: **Zytel® nylon resin replaced aluminum engine components under the hoods of dozens of automobile models to save weight.**

RIGHT BOTTOM: **The molded-in-color plastic fascia of the 2000 Dodge and Plymouth Neon was made with Surlyn® Reflection Series™ supergloss alloy.**

DuPont once again moved quickly, forming an alliance with General Foods to develop and improve soy foods. The company also announced a five-year research collaboration with the Massachusetts Institute of Technology to develop biotech applications outside agriculture. What Ed Jefferson had sensed in the potentials of chemistry, energy and the life sciences had become even clearer during Woolard's tenure as the company's chief environmentalist: there were fundamental connections among the many branches of science; between science and DuPont's historic research and business strengths; between those strengths and human needs; and between human needs and environmental integrity. Now, at the end of the company's second century, Chad Holliday summed it all up in a single vision for the company's future — sustainable growth. A sustainable growth company builds value for shareholders and society while decreasing its environmental footprint.

Holliday's November 1999 speech before the Economic Club of Detroit recalled Woolard's remarks before that body in 1990, when sustainable development was still a vague ideal. Woolard had worked to realize the environmental component of that ideal at DuPont by committing the company to specific, numerical goals. Now Holliday also defined sustainable growth by using measurable goals. By 2010, he predicted, DuPont would fill 10 percent of its energy needs and derive 25 percent of its income from renewable resources while delivering, as DuPont's new tagline put it, "The miracles of science®" to people around the world.[21]

Holliday's appointment in January 2000 as chair of the World Business Council for Sustainable Development underscored DuPont's global commitment to responsible, long-term growth. In a speech at the United Nations in April 2001, Holliday described three elements in DuPont's strategy to fulfill that commitment: integrated science, knowledge intensity and productivity. The first element, integrated science, stemmed from the basic connection between chemistry and biology that Jefferson had seen 20 years earlier. DuPont's new

655
DU PONT
The miracles of science®

DUPONT
AUTOMOTIVE
FINISHES
Fritos
QUAKER STATE
NOMEX
DELPHI
Bud
24
DUPONT
101
Santa Maria
Lompoc
Santa Ynez
Santa Barbara
N
50 miles
5
Ventura
Los Angeles
Pacific Ocean

Biomax® hydrobiodegradable packaging material, commercialized in 1999, was one example. Its new 3GT polyester technology, Sorona™, was another. DuPont's Kinston, North Carolina, plant was already producing a specialized polyester from petroleum feedstocks. But the same material could also be produced cheaply through the use of renewable carbohydrates and the patented *E. coli* bacteria DuPont had developed with Genencor. The Kinston plant prepared to start commercial production of Sorona™ in 2003. Bringing the Sorona™ bio-based technology on line was predicted to take at least seven years — about the same time it had taken to bring nylon from Wallace Carothers's lab to Wilmington's department stores in the 1930s. Successful commercialization still required patience, perseverance and plain perspiration.

In Holliday's understanding, integrated science was not limited to biology and chemistry. It also included any useful combination of science, engineering and technology, as in the Artistri® textile printing system. Introduced in 2000, Artistri® combined the company's long history in textiles with its more recent ink and ink jet printing experience to create a new, digital textile-printing technology for fabrics up to 10.5 feet wide. DuPont also formed a fuel cell business that year, applying its knowledge of electrochemistry, polymers and coatings to a single enterprise. Fuel cells stored electricity, like traditional batteries, but they recharged chemically, not electrically, using hydrogen and oxygen. Already appearing in "hybrid" cars, fuel cells offered an alternative to dependence on fossil fuels for a variety of energy needs.

DuPont's acquisition of a California company, UNIAX Corporation, in 2000 offered another product linking chemistry to energy savings. UNIAX had introduced a new display technology, like those used in watches, cell phones and laptop computers. Instead of liquid crystals, however, the lighter, more energy-efficient displays used a remarkable new discovery, a light-emitting polymer. Just seven months after DuPont's purchase in March, University of California chemist

BACKGROUND: **Drawing on pigment technology developed by DuPont scientist Harry Spinelli, Artistri® technology for digital textile printing was introduced in January 2001 by the DuPont Ink Jet business.**

NEAR LEFT: **DuPont has been the primary sponsor of Hendrick Motorsports' Car #24 driven by Jeff Gordon since he entered Winston Cup racing in 1993. Gordon has won four Winston Cup championships wearing a driver's suit of Nomex® and a helmet with Kevlar® sitting in cars painted with DuPont automotive finishes. Numerous DuPont products ride inside the car.**

FAR LEFT: **Sorona™, the company's newest polymer platform, will include an ingredient produced using a fermentation process based on corn sugar. Uncut staple material — displayed by Harold Thomas (left) and Bob Beckwith — was produced on a continuous polymerization unit in Kinston, North Carolina.**

BOTTOM: **Personal digital assistant displays developed by UNIAX Corporation are based on electroluminescent polymers. DuPont acquired UNIAX in 2000.**

CENTER: **In 1994, DuPont Nonwovens developed a soft cover of Tyvek® Plus that resists water and damaging ultraviolet rays.**

TOP RIGHT: **In Americana, Brazil, Raquel Aparecida Borges inspects nylon.**

BOTTOM RIGHT: **DuPont Protective Apparel — formed in 1997 and expanded in 1999 — combines the strengths of Tychem® chemical protective fabrics, Tyvek®, Kevlar®, Nomex® and Sontara® in one central collection of protective apparel knowledge.**

Alan Heeger, whose research had launched UNIAX in 1990, won the Nobel Prize for chemistry for his work on electroluminescent polymers.

The second element in DuPont's strategy, knowledge intensity, also was a new version of an older idea. For many years the company had marketed its expertise in areas like engineering, industrial safety and wastewater treatment. DuPont had "intensified" its knowledge to find new applications. In 1994, for instance, employees in DuPont's spunbonded fibers business had pooled their talents to design, produce and market a soft Tyvek® Plus automobile cover in less than a year. Holliday reinforced such efforts to capitalize on the company's assets. In 2000 DuPont realized new potential in several products, like Kevlar® and Nomex® fibers, Sontara® spunlaced fabric and Tychem® chemical protective fabrics, by combining its experience with these materials and its knowledge of safety issues to create the DuPont Protective Apparel Marketing Company. Now agencies such as fire and police departments and the chemical, electrical and waste disposal industries could look to DuPont for full-service advice on protecting workers. DuPont's Intellectual Assets Business also served to maximize the value of DuPont's many innovations, assisting the company's various businesses in licensing and selling DuPont's 17,000 active worldwide patents.

Productivity was the third part of DuPont's strategy to achieve sustainable growth. Here again the company was not discovering a new standard but fitting a traditional one to new conditions. In the 1990s these conditions included sharpened international competition; leaner, more market-focused businesses within the larger company; greater personal accountability and shared risk-taking for employees; and a faster pace of change driven by new communications technologies. Woolard had once observed that in a technically oriented company like DuPont, numbers often got more attention than words.[22] Holliday's adoption of the Six Sigma efficiency model put statistics on productivity, just as Woolard had put

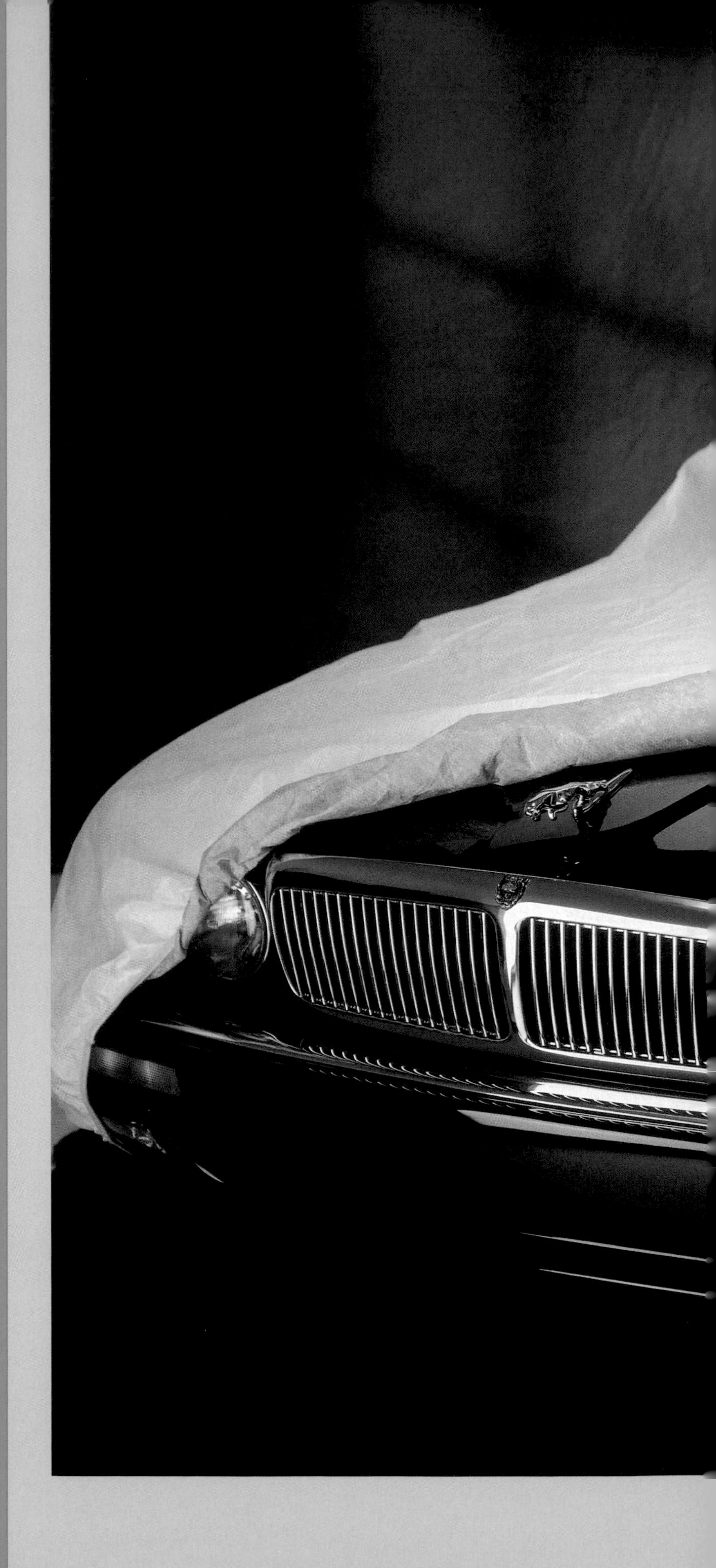

NEAR RIGHT: **DuPont introduced Zodiaq® quartz surfaces in 2000. Zodiaq® — composed of 93 percent quartz — is a new category of surfacing material for commercial and residential interiors.**

FAR RIGHT: **Timex timepieces use DuPont Holographics reflectors to increase display brightness, eliminate glare and heighten contrast.**

numbers on environmental goals, and as the entire DuPont Company had quantified workplace safety for decades. Zero defects now joined zero emissions and zero accidents among the goals DuPont set for itself.

The search for greater productivity also involved the continued restructuring of DuPont's global businesses. In April 2001 DuPont announced shutdowns at several of its older, less competitive polyester and nylon plants, and three months later it sold some polyester businesses to Alpek, part of a Mexican conglomerate. In June DuPont made a major decision to sell the pharmaceutical business to Bristol-Meyers Squibb Company, subject to government approval, for $7.8 billion, while retaining an interest in its Cozaar® and Hyzaar® antihypertensives. The pharmaceutical business had demanded massive, high-risk investments. DuPont was not shy about risk, but the stakes in pharmaceuticals were extraordinary and the business would require a disproportionate amount of funds relative to DuPont's other businesses. In theory, and often in practice, pharmaceuticals had been fertile ground for the kind of integrated science at which DuPont excelled, and the venture had brought some outstanding products to patients. The decision to sell DuPont Pharmaceuticals was not an easy one, but astute investors had always admired the company's willingness to eye its assets coolly and to shed them if necessary.[23] After 32 years in the business, DuPont set pharmaceuticals aside.

Balancing the risks and rewards of long-term research investment continued to challenge DuPont's leadership. In the early 1990s Chief Technology Officer Al MacLachlan had nurtured the vital connections between the company's businesses and its central research activities, giving all participants an important sense of shared mission and trust. His successor, Joseph A. Miller, Jr., sought to capitalize on that achievement by boosting DuPont's central research, thereby increasing the company's potential for major, breakthrough discoveries. The result was the Apex research process, introduced in 1998. In this process, a board of senior business leaders from DuPont evaluated long-range research proposals for their commercial potential, then provided initial corporate funding. As a project's likelihood of success became clearer, its funding would shift proportionally from corporate or central sources to a relevant DuPont business.[24] The Apex process affirmed the value of long-term risk-taking in a research-based enterprise and pointed to the importance of sharing the risks. The rigors of the 1990s had emphasized as never before that all DuPont employees were stakeholders and that all must be risk-takers as well. No amount of foresight or monitoring could remove the nail-biting uncertainty, the hopeful expectancy or the thrill of success that scientific research brought to a business.

By 2001, DuPont's businesses were organized around the linkages the company had established among its four major "technology platforms" — chemistry, biology, electronics and polymers — and its major markets, including transportation, construction, chemicals, textiles, health, electronics, and food and nutrition. The business logic guiding this global enterprise was clear. But DuPonters spontaneously expressed a strikingly uniform opinion about what held it all together. The company's technology was superb, its research was first-rate, its products were successful; however, it was DuPont's values and belief in its people that had made these things possible. DuPont was a good company, sure enough, but it was also good company — the kind they wished to keep.

"This company is special," Chad Holliday noted, "but it's special only if we keep it that way. It doesn't have any destiny that it *has* to be special." Holliday was convinced that DuPont and its people, each working hard to realize the best in the other, remained the surest formula for success. "If we're doing that right," he said, "we'll make the products."[25] The formula risked sounding like a corporate cliché, but Holliday and others used it anyway because they meant it and because there was plenty of evidence — 200 years of it — to back them up.

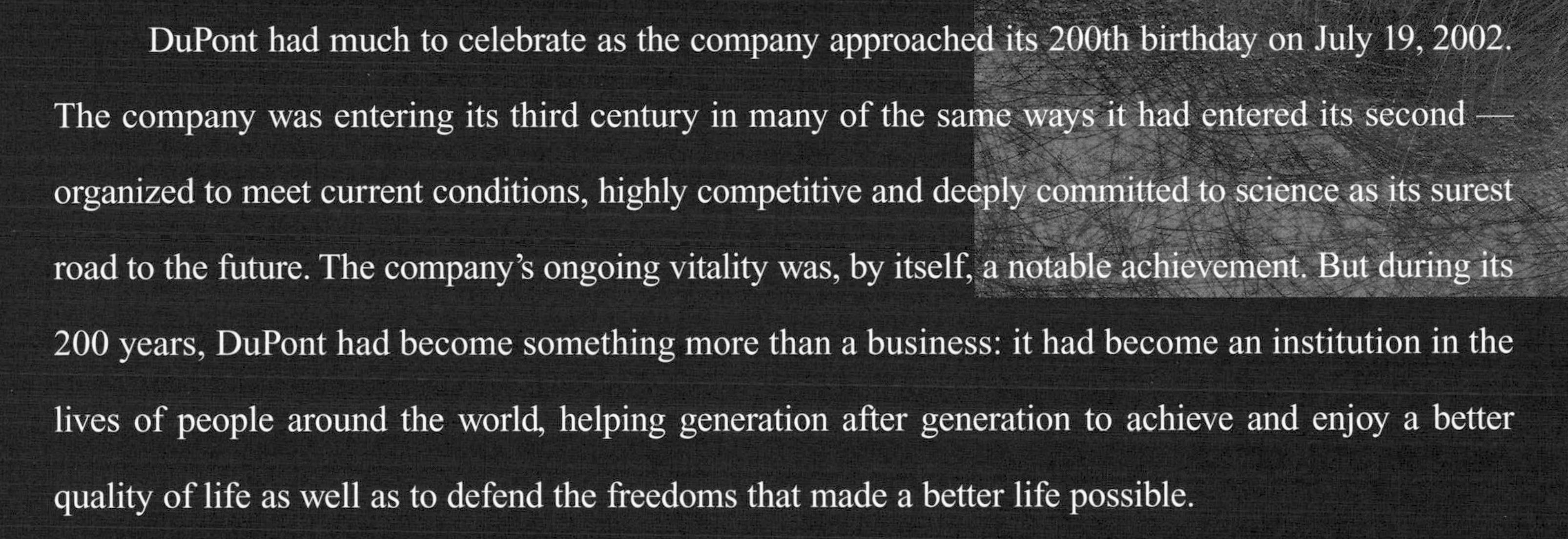

DuPont had much to celebrate as the company approached its 200th birthday on July 19, 2002. The company was entering its third century in many of the same ways it had entered its second — organized to meet current conditions, highly competitive and deeply committed to science as its surest road to the future. The company's ongoing vitality was, by itself, a notable achievement. But during its 200 years, DuPont had become something more than a business: it had become an institution in the lives of people around the world, helping generation after generation to achieve and enjoy a better quality of life as well as to defend the freedoms that made a better life possible.

In the last decades of the 20th century, DuPont's work extended far beyond America's shores, with plants and laboratories in 70 nations and products tailored to serve diverse markets in every corner of the world. The company's wide international experience contributed greatly to its keen vision of the future: environmentally sound growth in free markets for the betterment of all humankind. But visions themselves need sustaining. They thrive on a rare diet of patience, courage and, ultimately, success. DuPont has grown young on that diet for 200 years. It expects to grow younger still.

CHAPTER 1

1. For the du Pont family's origins in France and their roles in the Revolution, see William S. Dutton, *Du Pont: One Hundred and Forty Years* (New York: Charles Scribner's Sons, 1942), chapters 1–3. See also Joseph Frazier Wall, *Alfred I. du Pont: The Man and His Family* (New York: Oxford University Press, 1990).

2. Betty-Bright P. Low, *France Views America*, 1765–1815 (Wilmington, Del.: Eleutherian Mills Historical Library, 1978).

3. E. I. du Pont and Louis de Tousard had probably discussed the feasibility of a du Pont powder works prior to their much-celebrated misfires on a hunting excursion in the autumn of 1800, which reminded them of what they already knew about American-made powder.

4. Bessie Gardner du Pont, ed. and trans., *Life of Eleuthère Irénée du Pont From Contemporary Correspondence*, vols. 1–12 (Newark: University of Delaware Press, 1923–27), vol. 5, 198–213 (quote on 200), Microfiche, Library of American Civilization, LAC 20871–20876 (quote on 20873).

5. William C. Lawton, *The Du Ponts: A Case Study of Kinship in the Business Organization* (Ph.D. diss., University of Chicago, 1955), 94–96.

6. Louis de Tousard to E. I. du Pont, January 2, 1802, in *Life of Eleuthère Irénée du Pont*, 5:336.

7. Although the "d" in du Pont was capitalized in 1808, the space in "Du Pont" remained until 1992, when the official name of the company became "DuPont." The current spelling of the company name will be used hereafter.

8. *Papers of Eleuthère Irénée du Pont*, Longwood Manuscripts, Group 3, 146, file 30, Hagley Museum and Library, Greenville, Del. (HML).

9. Eleuthère Irénée du Pont to Mr. Robin, New York, January 25, 1802, in *Life of Eleuthère Irénée du Pont*, 5:360.

10. John C. Rumm, *Mutual Interests: Managers and Workers at the Du Pont Company, 1902–1915* (Ph.D. diss., University of Delaware, 1989).

11. For a description of the role of women's relationships to evangelical religion and to one another, of employee-owner relations, and of early 19th-century industry generally in the Brandywine Valley, see Anthony F. C. Wallace, *Rockdale: The Growth of an American Village in the Early Industrial Revolution* (New York: Knopf, 1978).

12. E. I. thought that Bauduy's choice of this name was "an attempt on the part of Mr. B for deceiving the dealers" (E. I. du Pont to George Boggs & Co., March 10, 1819); *Papers of Eleuthère Irénée du Pont*, Longwood Manuscripts, Group 3, 446, 284, HML.

13. See Norman B. Wilkinson, "Brandywine Borrowings From European Technology," *Technology and Culture* 4, no.1 (Winter 1963): 1–13.

14. Charles Grier Sellers, *The Market Revolution: Jacksonian America 1815–1846* (New York: Oxford University Press, 1991). See also George Rogers Taylor, *The Transportation Revolution*, 1815–1866 (New York: Rinehart, 1951).

15. Quoted in Rumm, *Mutual Interests*, 112.

16. Quoted in William H. A. Carr, *The du Ponts of Delaware* (London: Frederick Muller Limited, 1965), 150.

17. At the time of Alfred's resignation, the company was struggling under a debt load of half a million dollars. See Norman B. Wilkinson, *Lammot DuPont and the American Explosives Industry*, 1850–1884 (Wilmington, Del.: University Press of Virginia, for the Eleutherian Mills–Hagley Foundation, 1984), 26. Wilkinson speculates that *Scientific American* lifted the statement and the statistic from the previous month's *Delaware Gazette* (June 18, 1850). See Wilkinson, 18.

CHAPTER 2

1. Harold B. Hancock and Norman B. Wilkinson, "A Manufacturer in Wartime: DuPont, 1860–1865," *Business History Review* 40, no. 2 (Summer 1966): 213–36, 226–27.

2. Prices and taxes were closely related. DuPont justified its increased prices in 1864 by pointing out to the Army Ordnance Bureau that the cost of saltpeter had risen 135 percent since 1861. Sulphur had gone up 80 percent, charcoal 50 percent, kegs and barrels 90 percent and labor 75 percent. See Hancock and Wilkinson, "A Manufacturer in Wartime," 224.

3. Lammot once wrote, "In 1835 both Uncle Alexis & Uncle Henry came home to the powder — Uncle Alexis in September and Uncle Henry in the Spring of 1835." Lammot du Pont, "Powder Made By E. I. du Pont & Co., 1803–1856," written in 1856, *Lammot du Pont Papers*, Series B, Technical Papers, Accession #384, Box 33, HML.

4. Pierre Gentieu, "Reminiscences of One of DuPont's Employees," *Eugene du Pont, Family Miscellany*, 6, Accession #207, HML.

5. J.W. Macklem, "Old Black Powder Days," *The DuPont Magazine* 21, nos. 8–9, Anniversary Number (1927): 11, 46 (quote on p. 46).

6. Ibid., 2.

7. Ibid., 7–8. See also J.P. Monigle and Norman B. Wilkinson, *Oral History with Miss Katharine Collison*, September 1954–January 1955, Oral History Files, Accession #2026, Box 1, HML. "He looked very severe," recalled Miss Collison, "but he had a terribly soft melting kind of heart" (p. 4).

8. Hancock and Wilkinson, "A Manufacturer in Wartime," 221.

9. Quoted in Hancock and Wilkinson, "A Manufacturer in Wartime," 229.

10. Lammot du Pont to Mary Belin du Pont, February 25, 1880, in *Lammot du Pont Papers*, Accession #1579, Box 4, HML.

11. See Robert Wiebe, *The Search For Order*, 1877–1920 (New York: Hill and Wang, 1967), and Richard Hofstadter, *Social Darwinism in American Thought*, with a new introduction by Eric Foner (Boston: Beacon Press, 1955; 1992).

12. The firms were the DuPont Company, the Hazard Company, Laflin & Rand, Oriental Powder Mills, the Austin Powder Company, the American Powder Company and the Miami Powder Company. The California Powder Works joined in 1875. See Norman B. Wilkinson, *Lammot du Pont and the American Explosives Industry*, 1850–1884 (Charlottesville: University of Virginia Press, 1984), 203–30.

13. See Olivier Zunz, *Making America Corporate*, 1870–1920 (Chicago: University of Chicago Press, 1990).

14. In 1869 the Smith & Rand Powder Company consolidated with the Laflin Powder Company, creating the Laflin & Rand Powder Company; quoted in Wilkinson, *Lammot du Pont and the American Explosives Industry*, 227.

15. See Arthur VanGelder and Hugo Schlatter, *History of the Explosives Industry in America* (New York: Columbia University Press, 1927), 342–48.

16. Monigle and Wilkinson, *Oral History With Katharine Collison*, 4.

17. John C. Rumm, *Mutual Interests: Managers and Workers at the Du Pont Company, 1902–1915* (Ph.D. diss., University of Delaware, 1989), 174–75. See also, Kerby A. Miller, *Emigrants and Exiles: Ireland and the Irish Exodus to North America* (New York: Oxford University Press, 1985).

18. J.P. Monigle and Norman B. Wilkinson, *Oral History With William H. Buchanan*, August 7, 1958, Oral History Files, Accession #2026, Box 1, HML, 6, 23.

19. The two major biographies of Alfred are Marquis James, *Alfred I. du Pont, The Family Rebel* (New York: Bobbs-Merrill, 1941), and Joseph Frazier Wall, *Alfred I. du Pont: The Man and His Family* (New York: Oxford University Press, 1990).

20. J.P. Monigle and Norman B. Wilkinson, *Oral History With William H. Buchanan*, 3.

CHAPTER 3

1. See Alfred D. Chandler Jr., and Stephen Salsbury, *Pierre S. du Pont and the Making of the Modern Corporation* (New York: Harper & Row, Publishers, 1971).

2. Quoted in Chandler and Salsbury, *Pierre S. du Pont*, 35.

3. Ibid., 11, 27.

4. Ibid., 52–53.

5. Ibid., quote on 66.

6. See David A. Hounshell and John Kelly Smith, Jr., *Science and Corporate Strategy: DuPont R&D, 1902–1980* (New York: Cambridge University Press, 1988).

7. T. Coleman du Pont to Hamilton Barksdale, July 29, 1903, *Records of E. I. du Pont de Nemours & Company*, Series II, Part 2, *Papers of T. Coleman du Pont*, Box 807, Folder 23, HML.

8. John C. Rumm, *Mutual Interests: Managers and Workers at the Du Pont Company*, 1902–1915 (Ph.D. diss., University of Delaware, 1989), 215.

9. Quoted in William H.A. Carr, *The du Ponts of Delaware* (London: Frederick Muller Limited, 1965), 230.

10. Alfred du Pont to Frank Connable, November 20, 1906, Accession #1599, Box 1, HML; quoted in Rumm, *Mutual Interests*, 254.

11. For a discussion of Americans' mixed feelings about big business, see Ellis W. Hawley, *The New Deal and the Problem of Monopoly: A Study in Economic Ambivalence* (Princeton, N.J.: Princeton University Press, 1966; repr., New York: Fordham University Press, 1995).

12. See Davis Dyer and David B. Sicilia, *Labors of a Modern Hercules: The Evolution of a Chemical Company* (Boston: Harvard Business School Press, 1990), 41–64.

13. See L.L.L. Golden, *Only By Public Consent: American Corporations Search for Favorable Opinion* (New York: Hawthorn Books, Inc., 1968), 247–50.

14. Chandler and Salsbury, *Pierre S. du Pont*, 337.

15. Joseph Frazier Wall, *Alfred I. du Pont: The Man and His Family* (New York: Oxford University Press, 1990), 341–342.

16. Williams Haynes, ed., *American Chemical Industry: The Chemical Companies*, Vol. VI (New York: D. Van Nostrand Company, Inc., 1949), 132. The chief of the British Munitions Board was General Headlam.

17. John K. Winkler, *The Du Pont Dynasty* (New York: Reynal & Hitchcock, 1935), 244. Also, Chandler and Salsbury, *Pierre S. du Pont*, 418.

18. Winkler, *The Du Pont Dynasty*, 246.

CHAPTER 4

1. Quoted in Alfred P. Sloane Jr., *My Years With General Motors*, ed. John McDonald with Catharine Stevens (New York: Doubleday & Co., 1963; Anchor Books, 1972), 15.

2. P. J. Wingate, *The Colorful DuPont Company* (Wilmington, Del.: Serendipity Press, 1982), 39.

3. David A. Hounshell and John Kenly Smith Jr., *Science and Corporate Strategy: DuPont R&D, 1902–1980* (Cambridge: Cambridge University Press, 1988), 85. For a review of the effect of World War I on the American dyestuffs industry, see Ludwig F. Haber, *The Chemical Industry, 1900–1930: International Growth and Technological Change* (Oxford: Clarendon Press, 1971), 184–246. See also Graham D. Taylor and Patricia E. Sudnik, *DuPont and the International Chemical Industry* (Boston: Twayne Publishers, 1984), 43–58, 75–90, 105–30.

4. Taylor and Sudnik, *DuPont and the International Chemical Industry*, 107–11.

5. C. Chester Ahlum, "Cooperation With the Engineering Department," May 5, 1920, *Records of E. I. du Pont de Nemours & Company*, Accession #1784, Box 18, HML.

6. Joseph Borkin, *The Crime and Punishment of I.G. Farben* (New York: The Free Press, 1978), 38–40.

7. Charles W. Cheape, *Strictly Business: Walter Carpenter at DuPont and General Motors* (Baltimore: Johns Hopkins University Press, 1995), 39–45.

8. Jasper E. Crane, "A Short History of the Arlington Company," May 1, 1945, DuPont Pamphlet, HML. Mr. Crane was an executive with Arlington who continued his career at DuPont, becoming a vice president and a member of the Executive Committee in 1929.

9. Stephen Fenichell, *Plastic: The Making of a Synthetic Century* (New York: HarperCollins Publishers, 1996), 119–24.

10. Charles M. Stine, "The Kinship of du Pont Products," *The DuPont Magazine*, 19, 3 (March 1925), 1–2, 15.

CHAPTER 5

1. Quoted in L. G. Wise and N. G. Fisher, "History, Activities, and Accomplishments of Fundamental Research in the Chemical Department of the DuPont Company, 1926–1939 Inclusive," *Records of E. I. du Pont de Nemours & Company*, Accession #1784, Box 21, HML, 1.

2. David A. Hounshell and John Kelly Smith, Jr., *Science and Corporate Strategy: DuPont R&D, 1902–1980* (New York: Cambridge University Press, 1988), 226.

3. Matthew E. Hermes, *Enough For One Lifetime: Wallace Carothers, Inventor of Nylon* (American Chemical Society and Chemical Heritage Foundation, 1996), 83.

4. Raymond B. Seymour, ed., *Pioneers in Polymer Science* (Boston: Kluwer Academic Publishers, 1989), 34.

5. Hugh K. Clark, "Neoprene," *Papers of E. I. du Pont de Nemours & Company*, Accession #1850, HML, 1.

6. John K. Smith, "Interview with Dr. Merlin Brubaker," September 27, 1982, Oral History Interviews, Accession #1878, HML, 19.

7. Oliver M. Hayden, "Reflections on the Early Development of Neoprene," May 1, 1978, *Papers of E. I. du Pont de Nemours & Company*, Accession #1850, Box 6, HML, 4, 6.

8. David H. Hounshell and John K. Smith, "Interview with Julian Hill," December 1, 1982, Oral History Interviews, Accession #1878, no. 21, HML, 26.

9. Wallace H. Carothers, "Memorandum for Dr. A.P. Tanberg: Early History of Polyamide Fibers," February 19, 1936, *Records of E. I. du Pont de Nemours & Company*, Accession #1784, Box 18, HML.

10. Hounshell and Smith, "Interview with Julian Hill," 26.

11. Quoted in Hermes, *Enough for One Lifetime*, 139.

12. Robert S. McElvaine, *The Great Depression: America*, 1929–1941 (New York: Times Books, 1993), 72.

13. E. I. du Pont de Nemours & Company, *Annual Report* 1931, 5.

14. E. I. du Pont de Nemours & Company, *Annual Report* 1932, 6.

15. Elmer K. Bolton, "DuPont Research," *Industrial and Engineering Chemistry* 37, no. 2 (February 1945), 107–15.

16. C. M. A. Stine, "The Rise of the Organic Chemical Industry in the United States," *Annual Report of the Board of Regents of the Smithsonian Institution*, Washington, D.C., 1940, 177–92.

17. Hounshell and Smith, "Interview with Julian Hill," 53.

18. Alfred D. Chandler Jr. et al., "Interview with Elmer K. Bolton," 1961, Accession #1689, Oral History Interviews, HML, 21.

19. David A. Hounshell and John K. Smith, "Second Interview with Crawford H. Greenewalt," November 8, 1982, Oral History Interviews, Accession #1878, HML, 6–7.

20. Letter to Jasper Crane, cited in Charles W. Cheape, *Strictly Business: Walter Carpenter at DuPont and General Motors* (Baltimore: Johns Hopkins University Press, 1995), 132.

21. Letter to Lammot du Pont, May 18, 1935, *Records of E. I. du Pont de Nemours & Company*, Accession #1662, Box 3, HML.

22. Roland Marchand, *Creating the Corporate Soul: The Rise of Public Relations and Corporate Imagery in American Big Business* (Berkeley: University of California Press, 1998), 219. See also William L. Bird Jr., *"Better Living": Advertising, Media, and the New Vocabulary of Business Leadership, 1935–1955* (Evanston, Ill.: Northwestern University Press, 1999), and L. L. L. Golden, *Only By Public Consent: American Corporations' Search for Favorable Opinion* (New York: Hawthorn Books, Inc., 1968).

23. Joseph Labovsky, "A Short Biography of Nylon," undated, Nylon file, Pictorial Collection, HML.

24. Chandler et al., "Interview with Elmer K. Bolton," 20.

25. Hounshell and Smith, *Science and Corporate Strategy,* 270.

26. Hounshell and Smith, "Second Interview with Crawford H. Greenewalt," 27.

CHAPTER 6

1. See Richard G. Hewlett and Oscar E. Anderson Jr., *The New World, 1939–1946*, vol. 1, *A History of the United States Atomic Energy Commission* (University Park: Pennsylvania State University Press, 1962); Richard Rhodes, *The Making of the Atomic Bomb* (New York: Simon & Schuster, 1986); K. D. Nichols, *The Road to Trinity* (New York: William Morrow and Company, 1987); and Rodney Carlisle, with Joan M. Zenzen, *Supplying the Nuclear Arsenal: American Production Reactors*, 1942–1992 (Baltimore: Johns Hopkins University Press, 1996).

2. S. L. Sanger with Robert W. Mull, *Hanford and the Bomb: An Oral History of World War II* (Seattle: Living History Press, 1989), 25.

3. Ibid., 26.

4. Arthur Compton remembered Stine's odds as "one chance in a hundred." Arthur Holly Compton, *Atomic Quest: A Personal Narrative* (New York: Oxford University Press, 1956), 133.

5. Harry Thayer, *Management of the Hanford Engineer Works in World War II: How the Corps, DuPont and the Metallurgical Laboratory Fast Tracked the Original Plutonium Works* (New York: American Society of Civil Engineers Press, 1996), 73.

6. Crawford Greenewalt, Manhattan Project Diary, vol. 2, *Records of E. I. du Pont de Nemours & Company*, Accession #1889, HML, 5.

7. Leslie R. Groves, *Now It Can Be Told: The Story of the Manhattan Project* (New York: De Capo Press, 1975), 91–92.

8. Historians have questioned whether or not the use of atomic bombs against Japan was militarily, psychologically, and/or diplomatically necessary to induce surrender. Circumscribing the Soviet Union's postwar influence in Japan as well as Germany probably influenced U.S. decisions, although saving American lives continues to be regarded as the primary motive for the bombs' use. Gar Alperovitz explores several viewpoints in *The Decision to Use the Atomic Bomb* (London: HarperCollins Publishers, 1995).

9. E. I. du Pont de Nemours & Company, "DuPont's Role in the National Security Program, 1940–1945," March 7, 1946, pamphlet, HML.

10. John Morton Blum, *V Was for Victory: Politics and American Culture during World War II* (New York: Harcourt Brace Jovanovich, 1976).

11. William H. Chafe, *The Unfinished Journey: America since World War II* (New York: Oxford University Press, 1986), 10.

12. Congressional Research Service, *Congress and the Nation: A Review of Government and Politics in the Postwar Years* (Washington, D.C.: Congressional Quarterly, Inc., 1965), 114a–131a.

13. E. I. du Pont de Nemours & Company, *Annual Report* 1947, 27.

14. Crawford Greenewalt, *The Uncommon Man: The Individual in the Organization* (New York: McGraw-Hill Book Company, Inc., 1959).

15. "How to Win at Research," *Fortune Magazine*, October 1950, 115.

16. David A. Hounshell and John K. Smith, "Interview with Lester Sinness," October 23, 1985, Oral History Interviews, Accession #1878, HML, 17.

17. Roy J. Plunkett, interview by James J. Bohning in New York City and Philadelphia, April 14 and May 27, 1986, Chemical Heritage Foundation, Philadelphia, PA., 13.

18. Anne Cooper Funderburg, "Making Teflon Stick," *Invention & Technology* (Summer 2000), 10–20.

19. Ibid.

20. "Du Pont Replies to Government Charges," *Chemical & Engineering News 27*, no. 30 (July 25, 1949), 2181.

21. Crawford H. Greenewalt, "A Businessman Looks at the Antitrust Laws," August 13, 1963, pamphlet, HML, 2.

CHAPTER 7

1. *Time Magazine*, November 27, 1964, 95.

2. Quoted in James T. Patterson, *Grand Expectations: The United States, 1945–1974* (New York: Oxford University Press, 1996), 531.

3. David A. Hounshell and John K. Smith, "Interview with Edwin A. Gee," November 11, 1985, Oral History Interviews, Accession #1878, HML, 14.

4. Ibid., 46.

5. D. Brearley to G. J. Prendergast Jr., "Development Department Diversification Program of the 1960s," July 21, 1976, *Records of E. I. du Pont de Nemours & Company*, Accession #1850, HML.

6. David A. Hounshell and John K. Smith, "Third Interview with Chaplin Tyler," October 20, 1982, Oral History Interviews, Accession #1878, HML, 37.

7. Irving S. Shapiro, interview by James J. Bohning and Bernadette R. McNulty, transcript, December 15, 1994, Chemical Heritage Foundation, Philadelphia, Pa., 24.

8. Quoted in *Chemical & Engineering News*, May 30, 1966, 21.

9. D. H. Dawson, "Discussion with General Managers of Company Performance and Organization, April–July, 1961," July 14, 1961, *Records of E. I. du Pont de Nemours & Company*, Accession #1814, Series I, "Papers of Crawford H. Greenewalt," Box 3, HML.

10. L. S. Sinness to L. du P. Copeland, "Major Problems of the Company," August 21, 1964, *Records of E. I. du Pont de Nemours & Company*, Accession #1404, "Papers of Lammot du Pont Copeland," Box 9, HML.

11. Edwin A. Gee, "New Venture Development in DuPont," September 21, 1970, *Records of E. I. du Pont de Nemours & Company*, Accession #2232, Series II, "Papers of Edwin A. Gee," Box 2, HML, 1.
12. David A. Hounshell and John K. Smith, "Interview with Dr. Frank C. McGrew," August 2, 1983, Oral History Interviews, Accession #1878, HML, 53.
13. Howard E. Simmons Jr., interview by James J. Bohning, transcript, April 27, 1993, Chemical Heritage Foundation, Philadelphia, Pa., 21; Stephanie L. Kwolek, interview by Bernadette Bensaude-Vincent, transcript, March 21, 1998, Chemical Heritage Foundation, Philadelphia, Pa., 3–4.
14. L. du Pont Copeland, "Introductory Remarks, Wall St. Journal Visit," November 9, 1967, *Records of E. I. du Pont de Nemours & Company*, Accession #1404, "Papers of Lammot du Pont Copeland," Box 17, HML.
15. E. I. du Pont de Nemours & Company, *Annual Report* 1968, 31.
16. Joel A. Tarr, "Historical Perspectives on Hazardous Wastes in the United States," *Waste Management & Research 3* (1985), 95–102.
17. Dewey W. Grentham, *Recent America: The United States Since 1945* (Arlington Heights, Ill.: Harlan Davidson, Inc., 1987), 354.
18. E. I. du Pont de Nemours & Company, *Annual Report* 1973, 1.

CHAPTER 8

1. Harold M. Williams and Irving S. Shapiro, *Power and Accountability: The Changing Role of the Corporate Board of Directors* (Pittsburgh: Carnegie-Mellon University Press, 1979), 56.
2. Irving S. Shapiro, with Carl B. Kaufmann, *America's Third Revolution: Public Interest and the Private Role* (New York: Harper & Row, Publishers, 1984), 194–97.
3. James L. Phelan and Robert Pozen, *The Company State* (New York: Grossman, 1973).
4. Shapiro, *America's Third Revolution*, 71, 92.
5. Ibid., 46–47. Also, Thomas Byrne Edsall, *The New Politics of Inequality* (New York: W.W. Norton & Co., 1984), 155–57.
6. Howard E. Simmons, interview by James J. Bohning, DuPont Experimental Station, Wilmington, Del., April 27, 1993, Chemical Heritage Foundation, Philadelphia, Pa., 41, 48.
7. Shapiro, *America's Third Revolution*, 45.
8. William G. Simeral, telephone interview by Adrian Kinnane, July 18, 2001.
9. Harold C. Barnett, *Toxic Debts and the Superfund Dilemma* (Chapel Hill: University of North Carolina Press, 1994), 67–71.
10. Simeral telephone interview.
11. Chemical companies' unity in regard to passage of the CERCLA bill does not imply that they were united in their degree of support, or in their approach to environmental issues. See Paul Weaver, *The Suicidal Corporation* (New York: Simon and Schuster, 1988), 176–77. See also Ralph Nader and William Taylor, *The Big Boys: Power and Position in American Business* (New York: Pantheon Books, 1986), 186–89.
12. Godfrey Hodgson, *The World Turned Right Side Up: A History of the Conservative Ascendancy in America* (Boston: Houghton Mifflin Company, 1996), 193.
13. Edgar M. Bronfman, *Good Spirits: The Making of a Businessman* (New York: G.P. Putnam's Sons, 1998), 15.
14. *Wilmington News Journal*, June 12, 1982, A-3.
15. E. I. du Pont de Nemours & Company, *Annual Report* 1984, 2.
16. Francis Crick, *Of Molecules and Men* (Seattle: University of Washington Press, 1966).
17. Bronfman, *Good Spirits*, 167.
18. *Chemical & Engineering News* 67, no. 41 (October 9, 1989), 10–11.
19. *Chemical & Engineering News* 67, no. 41 (October 9, 1989), 12.
20. Ibid.
21. *Washington Post*, March 8, 1990, A-22.

CHAPTER 9

1. Edgar S. Woolard Jr., interview by Adrian Kinnane, Wilmington, Del., August 9, 2001. See also Edgar S. Woolard Jr., interview by James G. Traynham, June 10, 1999, Chemical Heritage Foundation, Philadelphia, Pa., 24.
2. Woolard interview by Traynham, 25.
3. Stacey J. Mobley, interview by John R. Rumm, DuPont Headquarters Building, Wilmington, Del., July 10, 2000, 33–34.
4. Robert J. Samuelson, "R.I.P.: The Good Corporation," *Newsweek*, 5 July 1993, 41.
5. E. I. du Pont de Nemours & Company, *Annual Report* 1993, 5.
6. Alexander MacLachlan, interview by Adrian Kinnane, Greenville, Del., July 2, 2001.
7. Ibid.
8. Edgar S. Woolard Jr., "Environmental Stewardship," *Chemical & Engineering News* 67, no. 22 (May 29, 1989),15; originally presented as a speech at the American Chamber of Commerce (U.K.), London, England, May 4, 1989.
9. Woolard, "Environmental Stewardship," 13.
10. Edgar S. Woolard Jr., "Creating Corporate Environmental Change," *The Bridge* (National Academy of Engineering) 29, no. 1 (Spring 1999); originally presented as a speech at the National Academy of Engineering, International Conference on Environmental Performance Metrics, Irvine, Cal., November 3, 1998.
11. Woolard interview by Traynham, 32.
12. Woolard interview by Kinnane.
13. Edgar S. Woolard, "Creating Corporate Environmental Change."
14. Woolard interview by Kinnane.
15. Ibid.
16. Bruce Herzog, DuPont Corporate Plans, interview by Adrian Kinnane, Wilmington Del., July 2, 2001.
17. On June 7, 2000, Optimum Quality Grains, LLC, became DuPont Specialty Grains following DuPont's 100 percent acquisition of Pioneer Hi-Bred International, Inc., in 1999.
18. Mobley interview, 25.
19. David Barboza, "DuPont Buying Top Supplier of Farm Seed," *New York Times*, March 16, 1999, C-2.
20. Charles O. Holliday, "Industry and Sustainability: Why should anyone take us seriously?" Speech before the Economic Club of Detroit, November 29, 1999.
21. Woolard, "Creating Corporate Environmental Change."
22. Marvin C. Brooks, personal communication, May 28, 2001. Brooks started his 40-year career with the U.S. Rubber Company, later UniRoyal, as a research chemist, then moved into national and international marketing with that company before retiring in the early 1980s.
23. MacLachlan interview.
24. Joseph A. Miller Jr., telephone interview by Adrian Kinnane, August 13, 2001.

Note: Page references in *italics* refer to illustrations.